T0205590

Access Control Models and Architectures For IoT and Cyber Physical Systems

Maanak Gupta • Smriti Bhatt
Asma Hassan Alshehri • Ravi Sandhu

Access Control Models and Architectures For IoT and Cyber Physical Systems

 Springer

Maanak Gupta
Department of Computer Science
Tennessee Technological University
Cookeville, TN, USA

Smriti Bhatt
Department of Computer and Information
Technology
Purdue University
West Lafayette, IN, USA

Asma Hassan Alshehri
College of Science and Humanities -
Dhurma
Shaqra University
Shaqra, Saudi Arabia

Ravi Sandhu
Institute for Cyber Security
University of Texas at San Antonio
San Antonio, TX, USA

ISBN 978-3-030-81091-7   ISBN 978-3-030-81089-4   (eBook)
https://doi.org/10.1007/978-3-030-81089-4

This Springer imprint is published by the registered company Springer Nature Switzerland AG
The registered company address is: Gewerbestrasse 11, 6330 Cham, Switzerland

Preface

Internet of Things (IoT) and Cyber Physical Systems (CPS) present the integration of the physical and digital worlds, offering a range of intelligent applications from manufacturing, transportation, energy, medical, buildings, and beyond. These promising paradigms, often used interchangeably, integrate multitude of heterogeneous, mobile, and dispersed physical smart devices which gather data from the environment and react to the needs offering automated experience to the end user. These systems incorporate entities and components from variety of sources that specify communication protocols, network requirements, physical resources characteristics, and real-time sensing and control using the capabilities of cloud, edge, or hybrid third-party services. CPS entails a distributed and dynamic ecosystem built with the Internet of Things as its backbone. These systems are becoming increasingly data rich, enabling new, unforeseen, and greater degrees of automation and autonomy to make our lives smarter.

Security and privacy are critical aspects when it comes to the wider deployment and adoption of CPS and IoT systems. These systems are subject to increasing threats and attacks from adversaries including state-sponsored and foreign-based entities. Cybersecurity threats exploit the proliferated distributed nature, complexity, physical and digital resources, multi domain, and connectivity of critical infrastructure systems, which are important for a nation's critical resources, putting them at serious risk. To cater these growing needs, it is important to identify and develop novel security mechanisms which are fundamentally different than the one usually used in single-domain enterprise like information computer systems.

Access control and authorization mechanisms have been widely used to limit and control unauthorized access to resources and assets of an organization to ensure confidentiality and integrity of the system. Conventional access control models including discretionary (DAC), mandatory (MAC), and role-based access control (RBAC) have been designed considering single-domain enterprise applications usually managed by a set of administrators and countable number of users in an organization. However, such models are difficult and expensive to adapt to distributed IoT and CPS systems, which are having thousands if not millions of smart connected devices that are managed across multiple domains within a single business unit. The dynamic and distributed behavior along with the introduction of multiple edge and cloud layers make it even harder to control resources and data which are gathered by them from the environment. We need to have a bigger picture of this connected ecosystem and develop a bottom-up approach with underlying fundamental technologies and frameworks to have a holistic solution to this cyber-elephant IoT and CPS security problem.

In this book, we will focus and present different conceptual, theoretical, and foundational access control solutions which have been developed by the authors to provide an overall picture and grounded approach to solve access control problems in IoT and CPS. We will present different architectures, frameworks, access control models, implementation scenarios, and broad set of use cases in different IoT and CPS domains to provide readers an intuitive and easy-to-read set of chapters. In addition, we will also discuss IoT and CPS access control solutions provided by key industry players including Amazon Web Services (AWS) and Google Cloud Platform (GCP), and discuss extensions of our proposed fine grained solutions with these widely used cloud and edge supported platforms.

The book is designed to serve the computer science and cybersecurity community including researchers, academicians, students, and practitioners who have wider interest in IoT/CPS privacy and security aspects. It provides a comprehensive document and recent development in terms of access control security of IoT and cyber physical systems. Thanks to the holistic planning and thoughtful organization of the book, the readers are expected to gain in-depth knowledge of the state-of-the-art access control architectures and security models for secure IoT and CPS.

We hope you enjoy reading this book!

Cookeville, TN, USA   Maanak Gupta
West Lafayette, IN, USA   Smriti Bhatt
Shaqra, Saudi Arabia   Asma Hassan Alshehri
San Antonio, TX, USA   Ravi Sandhu
January 2022

Acknowledgments

The authors are grateful to Farhan Patwa, associate director and chief architect, and James Benson, technology research associate II, Institute for Cyber Security, University of Texas at San Antonio, for their valuable suggestions and assistance in implementation of several technological solutions proposed in this work. In addition, we would also acknowledge all federal and private agencies including the US National Science Foundation (NSF), Department of Defense (DoD), and other organizations for providing grants for the research presented in the book. We are also grateful to numerous anonymous reviewers who provided useful comments which significantly improved the content published in this book.

Dr. Smriti Bhatt and Dr. Asma Hassan Alshehri have made equal contributions to this book and both should be jointly considered as second authors.

Contents

1 **Introduction: Requirements for Access Control in IoT and CPS** 1
  1.1   Introduction and Motivation 1
        1.1.1   IoT Architectures 2
        1.1.2   IoT and CPS Security Issues 4
  1.2   Access Control Models 6
        1.2.1   State of the Art 7
        1.2.2   Access Control Models for Smart Connected Systems 8
  1.3   Publish-Subscribe Paradigm 8
  1.4   IoT and CPS Integration with Cloud and Edge Computing 9
  1.5   Current Trends 10
  1.6   Access Control Challenges and Research Needs 12
  1.7   Summary 14
  References 14

2 **Access Control Oriented Architectures Supporting IoT and CPS** 19
  2.1   Introduction 19
        2.1.1   Chapter Organization 20
  2.2   Primitives for Cloud and Edge Assisted IoT 20
        2.2.1   Taxonomy of Smart Devices 20
        2.2.2   Cloud and Edge Hybrid Architectures 23
  2.3   Access Control Oriented Architectures 24
        2.3.1   Edge Gateway Supported ACO Architecture 28
        2.3.2   Extended ACO Architecture with Clustered Objects 29
  2.4   Illustrative IoT and CPS Using Proposed Architectures 32
        2.4.1   Remote Patient Monitoring (RPM) 32
        2.4.2   Intelligent Transportation System (ITS) 34
  2.5   Summary 36
  References 36

3 Authorization Frameworks for Smart and Connected Ecosystems   39
   3.1 Introduction   39
         3.1.1 Chapter Organization   40
   3.2 Access Control Framework for Cloud Enabled Wearable IoT   40
         3.2.1 Access Control Framework   41
         3.2.2 RPM Wearable IoT Use Case   44
   3.3 Framework for Smart Connected Cars Ecosystem   46
         3.3.1 Access Control Framework   48
         3.3.2 Identified Access Control Approaches........................   52
         3.3.3 Single and Multi-Cloud Cyber Physical Systems   54
   3.4 Objectives of Proposed Frameworks   58
   3.5 Summary   59
   References   60

4 Access Control Models in Cloud IoT Services   63
   4.1 Introduction   63
         4.1.1 Chapter Organization   64
   4.2 AWS Access Control Model..   64
   4.3 Access Control in AWS Internet of Things: AWS-IoTAC   67
         4.3.1 Motivation..   67
         4.3.2 Formal Model and Definitions   68
         4.3.3 AWS-IoTAC and ACO Architecture...........................   71
         4.3.4 Use Case   72
   4.4 Google Cloud Platform Access Control Model.......................   76
         4.4.1 GCP Access Control (GCPAC) Model   76
         4.4.2 Access Control in GCP Internet of Things   80
         4.4.3 E-Health Use Case...   84
   4.5 Limitations and Fine Grained Enhancements.........................   89
         4.5.1 Proposed Enhancements in AWS IoTAC   89
         4.5.2 Proposed Enhancements in GCP IoTAC......................   91
   4.6 Summary   94
   References   94

5 Secure Virtual Objects Communication   97
   5.1 Introduction   97
         5.1.1 Chapter Organization   98
   5.2 Operational Access Control for VO Communication   99
         5.2.1 ACL and Capability Based (ACL-Cap) Operational
                 Model..   99
         5.2.2 ABAC Operational Model.....................................   101
         5.2.3 RBAC Limitations   103
   5.3 Administrative Access Control for VO Communication   103
         5.3.1 Administrative ACL Model   104
         5.3.2 Administrative RBAC Model   105
         5.3.3 Administrative ABAC Model   106

5.4   AWS-IoT-ACMVO Model for AWS IoT Shadows
       Communication 108
5.5   Issues in Enforcing ACO-IoT-ACMsVO Within
       AWS-IoT-ACMVO 111
5.6   A Use Case: Sensing Speeding Cars 114
       5.6.1   Sensing the Speed of a Single Car 114
       5.6.2   Sensing the Speed of Multiple Cars 117
       5.6.3   Performance Evaluation and Discussion 120
5.7   Summary 121
References 122

6 **Attribute Based Access Control for Intelligent Transportation**......... 125
6.1   Introduction 125
       6.1.1   Chapter Organization 126
6.2   Authorization Requirements in ITS 127
       6.2.1   Multi-Layer and User Privacy Preferences 127
       6.2.2   Relevance of Groups 128
6.3   Dynamic Groups and ABAC Model 128
       6.3.1   CV-ABAC$_G$ Model Overview 129
       6.3.2   Components Definitions 134
6.4   AWS Enforcement 136
       6.4.1   Use Case Overview 136
       6.4.2   Prototype Implementation 137
       6.4.3   Performance Evaluation 142
6.5   Summary 144
References 144

7 **Fine Grained Communication Control for IoT and CPS** 147
7.1   Introduction 147
       7.1.1   Motivation 148
       7.1.2   Chapter Organization 149
7.2   Background and Related Work 150
       7.2.1   CE-IoT Architectures 150
       7.2.2   Related Work 150
       7.2.3   Scope and Assumptions 151
7.3   Access Control and Communication Control Requirements 152
       7.3.1   Use Case Scenarios 154
7.4   Attribute-Based Communication Control 155
       7.4.1   Attribute-Based Access Control Model 155
       7.4.2   Attribute-Based Communication Control Model 156
7.5   Attribute-Based Access and Communication Control Framework ... 159
       7.5.1   ABAC-CC Framework 159
7.6   Summary 161
References 162

8 Conclusions and Future Work 165
  8.1 Summary 165
    8.1.1 IoT and CPS Access Control Oriented Architectures 165
    8.1.2 Authorization Frameworks 166
    8.1.3 Access Control Formal Models 166
  8.2 Future Research Directions 167
    8.2.1 Communication Control in IoT and CPS 168

Index 171

Acronyms

ABAC	Attribute-Based Access Control
ABCC	Attribute-Based Communication Control
ABAC-CC	Attribute-Based Access and Communication Control
AC	Access Control
ACL	Access Control List
ACO	Access Control Oriented
AI	Artificial Intelligence
AP	Application
AWS	Amazon Web Services
AWSAC	Amazon Web Services Access Control
AWS-IoT	Amazon Web Services for Internet of Things
AWS-IoTAC	AWS-IoT Access Control
BSM	Basic Safety Message
CapBAC	Capability Based Access Control
CAN	Controller Area Network
CL	Cloud
CoAP	Constrained Application Protocol
CPS	Cyber Physical Systems
CO	Clustered Object
CV	Connected Vehicles
CE-IoT	Cloud-Enabled Internet of Things
CE-WIoT	Cloud-Enabled Wearable Internet of Things
CSR	Cloud Service
DAC	Discretionary Access Control
DDS	Data Distribution Service
DI	Direct Interaction
DL	Deep Learning
DMV	Department of Motor Vehicles
DSRC	Dedicated Short Range Communication
E-ACO	Extended Access Control Oriented
ECU	Electronic Control Unit

FG	FOG
GCP	Google Cloud Platform
GCPAC	Google Cloud Platform Access Control
GCP-IoTAC	Google Cloud Platform IoT Access Control
JSON	JavaScript Object Notation
HILF	High Impact and Low Frequency
HTTP	Hypertext Transfer Protocol
IdI	Indirect Interaction
IoT	Internet of Things
IoV	Internet of Vehicles
ITS	Intelligent Transportation System
M2M	Machine-to-Machine
MAC	Mandatory Access Control
ML	Machine Learning
MIoT	Medical IoT
MQTT	Message Queuing Telemetry Transport
OA	Object Abstraction
OB	Object
OBD	On Board Diagnostic
OAP	Object Layer Application
OTA	Over the Air
PKI	Public Key Infrastructure
RBAC	Role-Based Access Control
ReBAC	Relationship-Based Access Control
RPM	Remote Patient Monitoring
RFID	Radio Frequency Identification
SCMS	Security Credential and Management System
TCU	Telematic Control Unit
TPS	Tire Pressure Monitoring
U	User
USDOT	United States Department of Transportation
V2V	Vehicle-to-Vehicle
V2I	Vehicle-to-Infrastructure
V2P	Vehicle-to-Pedestrian
V2X	Vehicle-to-Everything
VIoT	Vehicular IoT
VO	Virtual Object
VOB	Virtual Object
VC	Vehicular Cloud
VCO	Vehicular Clustered Object
WIoT	Wearable IoT
WSN	Wireless Sensor Network

Chapter 1
Introduction: Requirements for Access Control in IoT and CPS

1.1 Introduction and Motivation

In the past decade, the notion of connected communities encompassing Internet of Things (IoT) and Cyber Physical Systems (CPS) has become an indispensable part of our lives. Several use cases have been developed in domains including transportation, building, manufacturing, energy grid, farming, healthcare etc. and have garnered attention from industry, academia and federal agencies. These CPS and IoT paradigms integrate geographically dispersed, distributed, heterogeneous smart devices such as sensors and actuators to collect large volumes of data, and to automate the entire connected ecosystem. This also integrates different technologies such as networking and communication protocols like WiFi, ZigBee, 5G, LoRA etc. together with cloud and edge computes needed for processing and storing humongous amounts of data. Several terminologies are used to refer this integration such as Cloud-Supported IoT, Cloud-Assisted IoT, and Cloud-Enabled IoT. We will also use this terminology interchangeably in this book.

The rapid growth of such connected ecosystem has resulted in infinite smart applications and has sparked potentially unrealistic expectations years ago. Several enterprises have started projects using technologies that integrate fundamental IoT architecture along with Artificial Intelligence (AI) and machine learning based applications. These IoT and CPS architectures merge the smart devices with dominant cloud services including Amazon Web Services, Google Cloud, Microsoft Azure etc. Such cloud enabled connected system can also be supported by edge based computational resources needed for domains require real time decision and reduce latency and bandwidth issues. However, most of the academic research has been focused in providing new applications and using *adhoc* cybersecurity solutions in IoT and CPS systems which were originally not designed to be lightweight and computationally inexpensive an absolute requirement for resource constrained environment. Because the IoT and CPS architecture encompasses pervasive connected heterogeneous objects, which interact with each other, with applications, and with other entities, security is necessary for its wide adoption and continued success. In addition, the integration of domains including cloud and edge-based computes broaden the attack surface, since it includes the risks at both the cloud/edge as well as the IoT vulnerabilities. The security and privacy of the data collected from personal and pervasive smart devices installed in the close vicinity, or sometimes on the user (like a smart watch or pacemaker) has the risk for privacy violation for both data at rest and in motion. The user data and information (e.g., their personal information and behavior patterns) gathered by these devices and shared with other components in these domains are highly privacy-sensitive.

In this chapter, we take a first step towards the goal of developing fine grained access control models for distributed cloud and edge assisted IoT and CPS domains. We first highlight the existing IoT architecture, which have been primarily divided into three layers: an object layer, one or more middle layers, and an application layer. We highlight the limitations of these architectures and highlight IoT security issues. We further provide some background knowledge of the widely used and established access control models used in enterprise applications, and some of the proposed solutions for IoT and CPS ecosystem. We also describe the widely deployed topic based lightweight publish subscribe based device communication, and how it is used to limit what data these devices can exchange, and what operations are allowed among devices. We also focus on the cloud and edge integration and the need for multi-cloud or edge cloud system to cater to different applications supporting real time or near real time solutions. In the last, we outline current trends and technological innovations with respect to authorization, and this leads us to discuss needed access control research to address some of the challenges or limitations of current solutions.

1.1.1 IoT Architectures

The main idea of the Internet of Things is to have remote objects that applications or other smart devices can communicate with and collect data from through the Internet. Therefore, there have been various proposals for IoT architecture and most of them have similar layers especially the top layer (application layer) and the bottom layer (object layer). Middle layer in the proposed architecture mostly is for objects communication, data transmission, and information processing.

Most initial proposed IoT architectures abstract the middle layer to be only one layer like network layer, whereas various recently proposed IoT architectures have two or more middle layers. Generally, the basic structure of IoT architecture is shown in Fig. 1.1, and it includes three main layers: object (lower) layer, one or more middle layers, and an application (upper) layer. The following are the main functionality and features of each layer [1].

- **The Object Layer:** Object layer in different proposed IoT architecture is called perception layer [2, 3] or hardware layer [4]. The purpose of the object layer

Fig. 1.1 The basic IoT
architecture

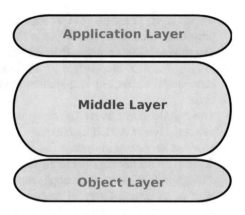

is to collect data from the surrounding environment of physical objects [5, 6], recognize objects [2], and rebuild a broad data visualization. Physical objects in the object layer can collect various data from environment including location, humidity, temperature, motion, speed, and direction [5, 7–10]. The main physical entity of the object layer is primarily sensors that collect data from nearby environment. Some proposed IoT architectures define structure of entities in the object layer as a network of only wireless sensors (e.g., cluster of sensors [11]), while other proposals beside sensors added actuators [4, 5], RFID tags [6], networks of devices (e.g., cameras and cellphones) [3, 12]. One of the most significant emerging technologies that produce big data is the IoT technology [13]. The IoT technology produce big data by depending on constrained pervasive and heterogeneous set of physical objects with different operating conditions, functionalities, resolutions, etc. [12]. The collected data can be non-structured or semi-structured data [14] because of the limited computational power and low storage of the physical objects in the object layer. These issues pose a great challenge and slow down the development of a unified reference model for the IoT [15]. Thus, the collected data needs to be manipulated and structured by upper layers with appropriate resources to deliver user functionality.

- **The Middle Layer:** This layer in the IoT architecture is mainly responsible about convey produced data in the object layer to remote destination through the Internet [3, 16]. Therefore, initial proposed IoT architecture define the middle layer as only one layer. In [3], authors referred to the middle layer as the transmission layer (gateway) which is in charge of handling communication between the application and object layer as well as gathering/sending data, packaging data, exchanging data, parsing/dispatching commands, and logging events. While in, [16–18], authors referred to the middle layer as the network layer which is responsible for intelligently processing the collected and produced big data. While some initial proposed IoT architectures assign only one major layer as the middle layer [3, 16], others referred to the middle layer with two or more extra layers between the application and object layer. In [19], the middle layer in the proposed IoT architecture includes the network layer and service layer. The network layer basically is for communication and transmission purposes, while the service layer is for services management and service APIs. The service layer is referred to middleware layer in the proposed IoT architecture in [20], which is responsible for information processing, link to the database, automatic decision, and a ubiquitous computation unit that can be placed in the cloud.

- **The Application Layer:** The computed and analyzed data is transferred to the most top layer in the IoT architecture. The application layer is the upper layer that provides services and system functionality to end users. The application layer depends on the middle layer to display collected and processed data to final users through user-friendly smart applications of the IoT technology. The application layer is an essential way to control and communicate remotely with physical objects/devices and display their information which can be used to generate models, graphs, and flowcharts. Decision makers use the analyzed data to assist them in decision-making process [2, 5, 21, 54].

1.1.2 IoT and CPS Security Issues

The IoT technology and its application are penetrating in many industries. In the IoT architecture, there are various entities in each layer of IoT architecture. As shown in Fig. 1.2, there are various entities of IoT which are basically users, objects (physical devices), and applications. The way of data transmission or entities communication could be within the same layer or different layers. However, exposing the collected data or unauthorized accessing to IoT entities is violating users' privacy and information security. Also, despite the supporting of IoT applications with the existing network technology, IoT applications still need to be secured with restrict security mechanisms and tools. Thus, the area of IoT security is one of the most significant factors that affect the development of the emerging IoT technology. In this section, the general IoT security issues and main threats are discussed upon the three layers on the general IoT architecture that is shown in Fig. 1.1.

- **Security Issues at Object layer:** The primary entity at object layer are sensors that sense the surrounding environment and collect data such as sound, location, temperature, and humidity. Also, actuator is another entity that execute a certain action on the physical environment. However, many of object layer entities have constrained capability in computation or storage. These entities can be vulnerable to a variety of attacks by the adversaries such as replacing the entity with a malicious one. Thus, new malicious entity appears to be part of the system and might expose confidential data [22]. Moreover, the memory of constrained entities in the object layer can be injected with malicious code that might cause entities to perform unauthorized actions or allow the attacker to gain access to the whole IoT system. Also, most of IoT entities are deployed within an open

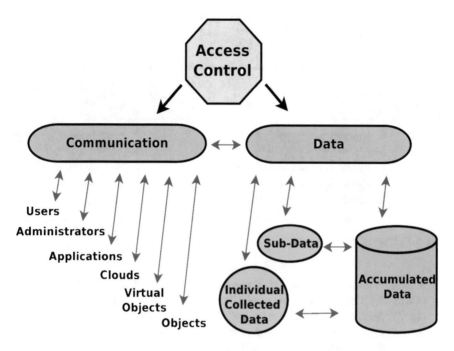

Fig. 1.2 Access control security issues in IoT and CPS

environment, so attackers will have the chance to eavesdrop confidential data in authentication or transmission phase [23].

- **Security Issues at Middle layer:** We stated in the general IoT architecture that middle layer is basically divided into two main layers; network layer and middleware layer. In the network layer, threats such as phishing attack are highly occurring. If user's account or password IoT system are compromised, the whole IoT/CPS system and environment becomes vulnerable to cyber attacks. Also because of the heterogeneity of CPS entities, complexity of IoT networks, and the weak configuration of IoT applications, the network layer is vulnerable to threats like Distributed Denial of Service (DDoS) that can be launched by attackers on the target servers by continuously spreading requests on poorly configured IoT entities [24]. Different threats like access attack, data transit attacks, and routing attack are likely to happen in the network layer [23, 25].

The middleware layer provides various services like information processing, link to the database, automatic decision, and a ubiquitous computation unit that can be placed in the cloud [20]. Also, in the middleware layer, there can be virtual [55] objects that represent physical objects in the object layer. Therefore, there are many crucial issues and questions can be addressed in the middle layer about virtual objects. How do virtual objects get permissions to access each other or to access the cloud? Are virtual objects permitted to communicate directly with cloud entities or not? And what are the conditions and requirements for this

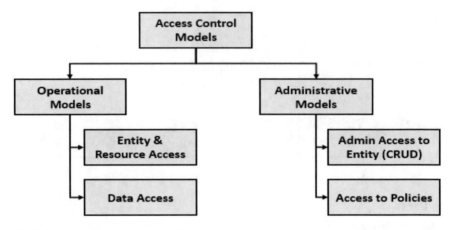

Fig. 1.3 Types of access control models

communication? Do clouds communicate to share their information with each other? Can virtual objects be controlled or accessed through different (remote) clouds? Such questions should be addressed with appropriate control.

- **Security Issues at Application layer:** The application layer is the top layer in the IoT architecture that presents collected and processed data to end users. In this layer, there are several related security issues and attacks can appear. Data theft is a critical threat that could happen especially with a lot of critical and private data that IoT applications can provide. Also, adversaries may use sniffer applications to observe the network traffic in the IoT applications, so unauthorized access to confidential user data can happen if there are no enough implemented security techniques [24]. Other possible threats like access control attack, malicious code injection attack, and reprogramming attack are likely to happen in the application layer [23]. Thus, developers of IoT and CPS applications need to apply security techniques and protocols such as data encryption, data isolation, user and network authentication, and privacy management against such threats.

1.2   Access Control Models

One of the main essential security technologies and mechanisms include authorization and access control. There are many access control approaches that have been proposed to control the access and grant authorization to subjects. Figure 1.3 shows different types of access control models. Access control models in general can be divided into two types: Operational and Administrative models. An operational access control model secures access on resources and services as well as their usage in any application or system. It also controls access to the data in a system.

Administrative access control model focuses on administrative accesses (e.g., create and delete) on various entities such as users, devices, objects/resources. The following are the main traditional access control models and some proposed access control model for distributed IoT and CPS systems.

1.2.1 State of the Art

Access control solutions have been extensively researched in the literature to offer information and resources security in computer and cyber systems. Many access control approaches have been proposed such as Access Matrix that express rights each subject owns for each object. Similarly, there is the access control list (ACL) that stores the access matrix by column, where each subject is associated with a capability list that store the matrix by row, and each object is associated with one or more subjects [26]. Other initial access control approaches are the discretionary access control (DAC) and mandatory access control (MAC). DAC permits owners to authorize each access request (e.g., read, write, or execute) of a user or group to objects. The access is granted if there exists an authorization stating that a user can make a specific access request to object. In spite of the flexibility of DAC, it is not restricted enough to enforce information flow policies since passing information from one object to another is not constrained. Conversely, MAC constrains information flow policies by assigning subjects and objects a security level (also called a clearance). Users have no choice to agree which users are permitted to access particular objects [26].

Role-based access control (RBAC) model [27–29] is one of the most commonly used approaches to control access and continue to be the preferred approach. With the RBAC model, administrators generate group of roles that express specific tasks, and assign list of permissions to the generated roles. This assignment called permission-role assignment are hard to change. Administrators later assign users to the generated roles to describe the authority and responsibility granted to a user. This assignment called user-role assignment change more frequently. Attribute-Based Access Control (ABAC) model [30–32] is an approach of access control that has lately attracted the attention of both academic and industry researchers. The US National Institute of Standards and Technology (NIST) described a high-level access control model with the attributes. Various access control approaches that use subject or object attributes have been proposed. The unified ABAC model that can be configured to the traditional access control models (i.e., DAC, MAC and RBAC) [30]. Also, the integration of the RBAC and ABAC approaches to control access has been studied by researchers. Bhatt et al. developed a role-centric access control model for OpenStack which integrated user attributes with RBAC [33].

1.2.2   Access Control Models for Smart Connected Systems

Many access control models for IoT and CPS have been proposed to address security and privacy issues, as surveyed in Ouaddah et al. [34]. Using capability-based access control (CAC) model for IoT has been proposed because entities hold granted rights that support different levels of granularity with possibility of delegation, while similar functionality is not feasible with ACLs. However, the main two major drawbacks of using the capability approach are propagation and revocation [27]. The identity authentication and capability-based access control (IACAC) model [35] is proposed, where devices use an access point and the CAC model to be connected to each other. Moreover, the capability-based access control system (CapBAC) is used in controlling access to services and information. The authors described use cases and argue that CapBAC supports rights delegation, least privileges access principle, more fine grained access control, fewer security issues, and fewer issues related to complexity and dynamics of subject's identities than ACLs, RBAC and ABAC. Bhatt et al. developed a formal access control model for AWS IoT platform and proposed ABAC enhancements for AWS IoT, a real world cloud-enabled IoT platform [36]. The authors also recently proposed a Convergent Access Control (CAC) framework that can converge access control features of different access control models (e.g., RBAC, ABAC, ReBAC, etc.) for enabling secure smart communities in the future [37]. Also, a simple-efficient mutual authentication and secure key establishment based on ECC, which has much lower storage and communication overheads, is proposed for the perception (object) layer of the IoT [34]. Gupta and Sandhu proposed a novel perspective by introducing activity centric access control [38] for smart collaborative systems, assuming activity aka task as the prime notion to control new activities in the connected CPS systems.

1.3   Publish-Subscribe Paradigm

Publish-Subscribe (Pub/Sub) paradigm allows to send and receive messages through topics or channels between devices, applications, and/or services which act as a publisher or subscriber. The publisher (source of data) can publish messages on topics/channels and a subscriber (receiver of data) can subscribe to topics/channels to receive messages published on specific subscribed topic/channels. The pub/sub communication paradigm is suitable for large-scale distributed interactions and has become a widely used communication paradigm in IoT and CPS today.

Figure 1.4 shows a Pub-Sub paradigm. In Pub/Sub paradigm, publishers publish messages with desired payloads on specific topics/channels and need not to be aware of the subscribers. Subscribers on the other hand need to subscribe to specific topics from which they intend to receive messages, and once subscribed, whenever there is a new message published on that topic, the message broadcasts to all the subscribers. An IoT device, application, or service can be a publisher, subscriber, or

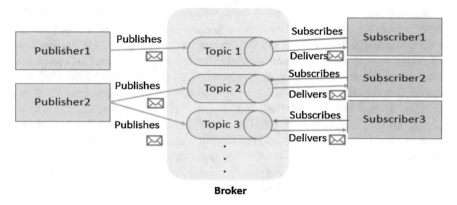

Fig. 1.4 A view of publish-subscribe paradigm

both based on the use case scenario. The message payload can be of any data type such as strings, JSON, text, audio, video, and other complex object types. Generally, there is a message broker that handles how the messages are distributed based on subscriptions and manages any access control associated, such as who can publish messages on a specific topic, and who is allowed to subscribe to those topics.

There are different types of implementation styles in Pub/Sub among which two commonly used are—*topic-based* and *content-based* [39, 40]. In the topic-based scheme, a topic or channel maps to a group and subscribers can become a member of a specific topic (a group), and publishers can publish messages to that topic. Pub/Sub is based on an event-driven communication mechanism and notification scheme. All subscribers to that topic are informed about the published data. Whereas, the content-based approach focuses on the content of the message and introduces a subscription scheme based on the published content. Here, subscribers specify filters, which define constraints based on the name-value pairs of the published properties (content) and use single or combined basic comparison operators ($=$, $<$, $\leq$, $>$, $\geq$) to identify events of interest [39, 40]. A widely used publish-subscribe paradigm in the industry is MQTT [41]. It is a standard messaging protocol for IoT and is being widely used various IoT domains today, such as automotive, manufacturing, telecommunications, oil and gas [42, 56], etc.

1.4   IoT and CPS Integration with Cloud and Edge Computing

Cloud computing is a well-established architecture today that IoT leverages for data storage, computation, and analytics. It has become a key enabling technology for IoT and data-driven applications that are connected to IoT devices and utilizes data collected from these devices. With billions of connected devices, cloud provides foundation services to support IoT devices, which are usually resource-constrained, along with IoT applications for sustainable success and wide adoption of IoT in real-world. The integration of cloud and IoT has been recently suggested in the literature [5, 14, 21, 43, 44]. It has also been adopted in the industry by major cloud computing services providers—Amazon Web Services (AWS) [45], Google Cloud [46], and Azure Cloud [47], to enable IoT services and applications empowered by smart devices. Several terminologies are used to refer this integration, such as Cloud-Supported IoT, Cloud-Assisted IoT [7], and Cloud-Enabled IoT [1], more widely used as Cloud-Enabled IoT (CE-IoT) [48, 49]. The integration of these two broad domains raise many security and privacy concerns, since a large attack surface, including both cloud computing and IoT vulnerabilities, is exposed to the attackers. Another significant security and privacy concern is associated with the vast amount of data and information continuously being generated, stored, and shared between IoT devices and different entities (e.g., users, applications, and services) in CE-IoT architectures. The IoT data comprises users' and/or organizations' data, which can be highly privacy sensitive and is at constant risk from malicious users and attackers.

The number of connected IoT devices is tremendously increasing, and according to Gartner, it has been expected that there will be more than 20 billion connected devices by 2020. With such as large number of connected devices, there is a huge amount of data associated and continuously being generated by them and not every bit of data is always useful or required to be sent or stored in the cloud. Therefore, edge computing is necessary for effective IoT where some amount of computation, storage, and analytics capabilities are moved towards the edge of the network. Thus, an edge-centric approach in addition to the cloud-centric approach for IoT is utilized to support evolving IoT application domains and for continued success of IoT in the future. Figure 1.5 shows a cloud and edge/fog computing enabled IoT vision. It is becoming a popular architecture to empower future IoT devices and applications. Edge computing also enabled local computation and communication when there is limited or intermittent connectivity with cloud providers. Therefore, it is essential to have cloud and edge computing enabled IoT architectures for enabling connected smart communities. In order to employ the concept of edge computing and distribute some of the cloud capabilities towards the edge of the network in CE-IoT, there is a need to utilize and deploy edge gateways or multiple small edge cloudlets [9, 48, 50].

1.5 Current Trends

IoT and CPS are rapidly evolving space which is enabled by other key technologies, such as cloud computing, edge/fog computing, data analytics, connectivity, and machine-to-machine (M2M) communication protocols. Recently, Artificial Intelligence (AI) and Machine Learning (ML) approaches are also largely being explored in the IoT domain, such as AI-enabled smart devices, robots, automation, and data-driven applications empowered by ML algorithms. While these depicts some of the current trends in IoT, there are many technological advancements within above areas questions such as, (a) What are the necessary requirements for objects to authorize them to communicate? (b) Which objects are authorized to communicate with a specific object? (c) Which virtual object can a device communicate or what applications can issue a command to the device? Even among virtual objects, which devices are allowed to publish to certain topics, which devices can subscribe to reserved topics, can all the data on the subscribed topic be shared, or any contextual conditions when the permission to communicate is allowed. Appropriate access control solutions are needed to control such inter layer communication. Further, as the number of devices and services grow, it is imperative for the device owners and the administrators to enforce access control policies without understanding what a particular entity may have access to in a different domain. Since administrator and users both access through applications, it is important to distinguish administrators from users. How communications between entities at same and different layers can be controlled? What operations that are controlled by individual and how they are different from actions that need direct control from administrators?

- **Controlled Data Access:** Data collected by the ubiquitous smart devices can be requested by other devices to make data driven decisions. In addition, there is meta-data related to different entities in IoT and CPS such as information about the devices, its virtual objects, cloud registry, and user applications which can access certain resources. Data exchange is the result of communication among devices; therefore, data security is needed at each stage of the data lifecycle since the likelihood of it getting leaked will exist until it is permanently deleted from central cloud or edge services. Confidentiality and integrity of this data must be maintained at every stage of the lifecycle. Several important questions like, if an object, application or virtual object can access data partially, and if so, can the data be retrieved across different entities. Another important concern is whether the data will be stored in single cloud, or across multiple cloud systems and how it will be accessed by entities associated with different cloud.
- **Multi Cloud and Edge Cloud Interactions:** Billions of connected devices will require the need for a hybrid cloud edge and multi cloud architecture. In addition, devices will be associated with different cloud providers and may need trusted collaboration among them before any information crosses their administrative domain. Using edge or fog computing is critical to resolve the issues of low bandwidth, high latency, and communication delays common in using central cloud, which are very important in location and time sensitive IoT and CPS applications. Cross cloud access controls and relevant security models are still at infancy stage and need more focused attention. Appropriate Trust-based access control models for cross-tenant, cross-account, and multi-cloud architectures are needed for domains like vehicular IoT, smart cars, grids and wearable IoT.
- **User Privacy Preferences:** Users are critical to the smart ecosystem and must be kept in the loop for making any informed access control decision not only with respect to the data collected from associated devices, but also who can control the devices the user is associated with, or what kind of alerts a user may want to receive on his or her car dashboard. For instance, a user may not want to

receive restaurant notifications but is interested in flash-flood warnings. A user must be given the option to set his personal preferences whether he wants to receive advertisements from a restaurant or filter out which ones are acceptable. We firmly believe that end user must be involved in authorization process, which may not be needed for each access control request but may have the flexibility to set preferences at the installation time and change it as needed. It may be needed to have multi-level security policy with system wide policies deployed by the device manufacturer or cloud provider, and secondary deployed by user for his/her preferences. Such requirements play important role when deployed in CPS domains including smart connected cars, or wearable IoT.

1.7   Summary

In this chapter, we discussed the introduction and need for novel access control models for IoT and cyber physical systems. We discussed the state-of-the-art access control and how they can be extended for such domains. We provided an overview of different IoT and CPS architectures, and briefly highlighted the publish subscribe topic-based device communication. We also elaborated on the current trends in authorization for cloud assisted IoT systems and discussed challenges and future research needs. In the following chapters, we will provide deeper understanding about the different access control architectures, develop formal security models in different CPS domains and discuss cloud and edge assisted fine grained access control models.

References

1. Alshehri, A., & Sandhu, R. (2016). Access control models for cloud-enabled internet of things: A proposed architecture and research agenda. In *2016 IEEE 2nd International Conference on Collaboration and Internet Computing (CIC)* (pp. 530–538). IEEE.
2. Wu, M., Lu, T.-J., Ling, F.-Y., Sun, J., & Du, H.-Y. (2010). Research on the architecture of internet of things. In *2010 3rd International Conference on Advanced Computer Theory and Engineering (ICACTE)* (vol. 5, pp. V5–484). IEEE.
3. Zhu, Q., Wang, R., Chen, Q., Liu, Y., & Qin, W. (2010). Iot gateway: Bridgingwireless sensor networks into internet of things. In *2010 IEEE/IFIP International Conference on Embedded and Ubiquitous Computing* (pp. 347–352). IEEE.
4. Gubbi, J., Buyya, R., Marusic, S., & Palaniswami, M. (2013). Internet of things (IoT): A vision, architectural elements, and future directions. *Future Generation Computer Systems, 29*(7), 1645–1660.
5. Al-Fuqaha, A., Guizani, M., Mohammadi, M., Aledhari, M., & Ayyash, M. (2015). Internet of things: A survey on enabling technologies, protocols, and applications. *IEEE Communications Surveys & Tutorials, 17*(4), 2347–2376.
6. Porambage, P., Ylianttila, M., Schmitt, C., Kumar, P., Gurtov, A., & Vasilakos, A. V. (2016). The quest for privacy in the internet of things. *IEEE Cloud Computing, 3*(2), 36–45.

7. Gupta, M., & Sandhu, R. (2018). Authorization framework for secure cloud assisted connected cars and vehicular internet of things. In *Proc. of the 23nd ACM on Symposium on Access Control Models and Technologies* (pp. 193–204).
8. Gupta, M., Benson, J., Patwa, F., & Sandhu, R. (2019). Dynamic groups and attribute-based access control for next-generation smart cars. In *Proceedings of the Ninth ACM Conference on Data and Application Security and Privacy* (pp. 61–72).
9. Gupta, M., Benson, J., Patwa, F., & Sandhu, R. (2020). Secure V2V and V2I communication in intelligent transportation using cloudlets. *IEEE Transactions on Services Computing.* https://doi.org/10.1109/TSC.2020.3025993.
10. Gupta, M., Abdelsalam, M., Khorsandroo, S., & Mittal, S. (2020a). Security and privacy in smart farming: Challenges and opportunities. *IEEE Access, 8,* 34564–34584.
11. Voas, J. (2016). Networks of 'things'. *NIST Special Publication, 800*(183), 800–183.
12. Sarkar, C., Uttama Nambi SN, A., Venkatesha Prasad, R., Rahim, A., Neisse, R., & Baldini, G. (2014). Diat: A scalable distributed architecture for IoT. *IEEE Internet of Things Journal, 2*(3), 230–239.
13. Dobre, C., & Xhafa, F. (2014). Intelligent services for big data science. *Future Generation Computer Systems, 37,* 267–281.
14. Botta, A., De Donato, W., Persico, V., & Pescapé, A. (2016). Integration of cloud computing and internet of things: A survey. *Future Generation Computer Systems, 56,* 684–700.
15. Nitti, M., Pilloni, V., Colistra, G., & Atzori, L. (2015). The virtual object as a major element of the internet of things: A survey. *IEEE Communications Surveys & Tutorials, 18*(2), 1228–1240.
16. Yang, Z., Yue, Y., Yang, Y., Peng, Y., Wang, X., & Liu, W. (2011). Study and application on the architecture and key technologies for IoT. In *2011 International Conference on Multimedia Technology* (pp. 747–751). IEEE.
17. Jia, X., Feng, Q., Fan, T., & Lei, Q. (2012). Rfid technology and its applications in internet of things (IoT). In *2012 2nd International Conference on Consumer Electronics, Communications and Networks (CECNet)* (pp. 1282–1285). IEEE.
18. Domingo, M. C. (2012). An overview of the internet of things for people with disabilities. *Journal of Network and Computer Applications, 35*(2), 584–596.
19. Da Xu, L., He, W., & Li, S. (2014). Internet of things in industries: A survey. *IEEE Transactions on Industrial Informatics, 10*(4), 2233–2243.
20. Khan, R., Khan, S. U., Zaheer, R., & Khan, S. (2012). Future internet: the internet of things architecture, possible applications and key challenges. In *2012 10th International Conference on Frontiers of Information Technology* (pp. 257–260). IEEE.
21. Atzori, L., Iera, A., & Morabito, G. (2010). The internet of things: A survey. *Computer Networks, 54*(15), 2787–2805.
22. Kumar, S., Sahoo, S., Mahapatra, A., Swain, A. K., & Mahapatra, K. K. (2017). Security enhancements to system on chip devices for IoT perception layer. In *2017 IEEE International Symposium on Nanoelectronic and Information Systems (iNIS)* (pp. 151–156). IEEE.
23. Hassija, V., Chamola, V., Saxena, V., Jain, D., Goyal, P., & Sikdar, B. (2019). A survey on IoT security: Application areas, security threats, and solution architectures. *IEEE Access, 7,* 82721–82743.
24. Kolias, C., Kambourakis, G., Stavrou, A., & Voas, J. (2017). Ddos in the IoT: Mirai and other botnets. *Computer, 50*(7), 80–84.
25. Sontowski, S. et al. (2020). Cyber attacks on smart farming infrastructure. In *Proc. of the IEEE Conference on Collaboration and Internet Computing (CIC).* https://doi.org/10.1109/CIC50333.2020.00025
26. Sandhu, R. S., & Samarati, P. (1994). Access control: principle and practice. *IEEE Communications Magazine, 32*(9), 40–48.
27. Sandhu, R. S., Coyne, E. J., Feinstein, H. L., & Role-Based, C. E. Y. (2013). Access control models. *IEEE Computer, 29*(2), 38–47.
28. Sandhu, R. S., Coyne, E. J., Feinstein, H. L., & Role-Based, C. E. Y. (1996). Role-based access control models yz. *IEEE Computer, 29*(2), 38–47.

29. Ferraiolo, D. F., Sandhu, R., Gavrila, S., Kuhn, D. R., & Chandramouli, R. (2001). Proposed nist standard for role-based access control. *ACM Transactions on Information and System Security (TISSEC), 4*(3), 224–274.

30. Jin, X., Krishnan, R., & Sandhu, R. (2012). A unified attribute-based access control model covering dac, mac and rbac. In *IFIP Annual Conference on Data and Applications Security and Privacy* (pp. 41–55). Springer.

31. Gupta, M., & Sandhu, R. (2016). The GURA$_G$ administrative model for user and group attribute assignment. In *International Conference on Network and System Security* (pp. 318–332). Springer.

32. Gupta, M., Patwa, F., & Sandhu, R. (2018). An attribute-based access control model for secure big data processing in hadoop ecosystem. In *Proceedings of the Third ACM Workshop on Attribute-Based Access Control* (pp. 13–24).

33. Bhatt, S., Patwa, F., & Sandhu, R. (2016). An attribute-based access control extension for openstack and its enforcement utilizing the policy machine. In *2016 IEEE 2nd International Conference on Collaboration and Internet Computing (CIC)* (pp. 37–45). IEEE.

34. Ouaddah, A., Mousannif, H., Abou Elkalam, A., & Ouahman, A. A. (2017). Access control in the internet of things: Big challenges and new opportunities. *Computer Networks, 112,* 237–262.

35. Mahalle, P. N., Anggorojati, B., Prasad, N. R., & Prasad, R. (2013). Identity authentication and capability based access control (iacac) for the internet of things. *Journal of Cyber Security and Mobility, 1*(4), 309–348.

36. Bhatt, S., Patwa, F., & Sandhu, R. (2017). Access control model for aws internet of things. In *International Conference on Network and System Security* (pp. 721–736). Springer.

37. Bhatt, S., & Sandhu, R. (2020). Convergent access control to enable secure smart communities. In *2020 Second IEEE International Conference on Trust, Privacy and Security in Intelligent Systems and Applications (TPS-ISA)* (pp. 148–156). IEEE.

38. Gupta, M., & Sandhu, R. (2021). Towards activity-centric access control for smart collaborative ecosystems. Preprint. arXiv:2102.11484.

39. Bacon, J., Eyers, D. M., Singh, J., & Pietzuch, P. R. (2008). Access control in publish/subscribe systems. In *Proceedings of the Second International Conference on Distributed Event-Based Systems* (pp. 23–34).

40. Eugster, P. T., Felber, P. A., Guerraoui, R., & Kermarrec, A.-M. (2003). The many faces of publish/subscribe. *ACM Computing Surveys (CSUR), 35*(2), 114–131.

41. MQTT: The Standard for IoT Messaging. https://mqtt.org/. [Online; accessed 20-January-2021]

42. Gupta, M., Awaysheh, F. M., Benson, J., Al Azab, M., Patwa, F., & Sandhu, R. (2020b). An attribute-based access control for cloud-enabled industrial smart vehicles. *IEEE Transactions on Industrial Informatics.* https://doi.org/10.1109/TII.2020.3022759.

43. Roman, R., Zhou, J., & Lopez, J. (2013). On the features and challenges of security and privacy in distributed internet of things. *Computer Networks, 57*(10), 2266–2279.

44. Prahlada Rao, B. B., Saluia, P., Sharma, N., Mittal, A., & Sharma, S. V. (2012). Cloud computing for internet of things & sensing based applications. In *2012 Sixth International Conference on Sensing Technology (ICST)* (pp. 374–380). IEEE.

45. Amazon Web Services. https://aws.amazon.com/. [Online; accessed 04-January-2021].

46. Google Cloud Platform. https://cloud.google.com/docs. [Online; accessed 10-December-2020].

47. Microsoft Azure. https://azure.microsoft.com/en-us/. [Online; accessed 24-December-2020].

48. Bhatt, S., Patwa, F., & Sandhu, R. (2017). An access control framework for cloud-enabled wearable internet of things. In *2017 IEEE 3rd International Conference on Collaboration and Internet Computing (CIC)* (pp. 328–338). IEEE.

49. Bhatt, S., Lo'ai, A. T., Chhetri, P., & Bhatt, P. (2019). Authorizations in cloud-based internet of things: current trends and use cases. In *2019 Fourth International Conference on Fog and Mobile Edge Computing (FMEC)* (pp. 241–246). IEEE.

50. Satyanarayanan, M., Bahl, P., Caceres, R., & Davies, N. (2009). The case for vm-based cloudlets in mobile computing. *IEEE pervasive Computing, 8*(4), 14–23.
51. California IoT Security Law Cheat Sheet. https://www.jdsupra.com/legalnews/california-iot-security-law-cheat-sheet-75568/. [Online; accessed 27-January-2021].
52. Gupta, M., Patwa, F., & Sandhu, R. (2017). Object-tagged rbac model for the hadoop ecosystem. In *IFIP Annual Conference on Data and Applications Security and Privacy* (pp. 63–81). Springer.
53. Bhatt, S., & Sandhu, R. (2020). Abac-cc: Attribute-based access control and communication control for internet of things. In *Proceedings of the 25th ACM Symposium on Access Control Models and Technologies* (pp. 203–212).
54. Gupta, M., Patwa, F., & Sandhu, R. (2017). POSTER: Access control model for the Hadoop ecosystem. In *Proceedings of the 22nd ACM on Symposium on Access Control Models and Technologies* (pp. 125–127).
55. Cathey, G., Benson, J., Gupta, M., & Sandhu, R. (2021). Edge centric secure data sharing with digital twins in smart ecosystems. *Preprint arXiv:2110.04691.*
56. Bhatt, S., Pham, T. K., Gupta, M., Benson, J., Park, J., & Sandhu, R. (2021). Attribute-based access control for AWS internet of things and secure Industries of the Future. *IEEE Access, 9,* 107200–107223.

Chapter 2
Access Control Oriented Architectures Supporting IoT and CPS

2.1 Introduction

This chapter offers a steppingstone for our eventual goal of developing an authoritative family of access control models for a cloud-enabled Internet of Things. We will build upon the IoT architecture which are all roughly divided into three layers: an object layer, one or more middle layers, and an application layer, as discussed in the previous chapter. The proposed access-control oriented (ACO) architecture supports cloud and virtual objects in the middle of object and application layers. This architecture is divided into four layers: *object* layer, *virtual object* layer, *cloud* layer, and *application* layer. Each of these layers encapsulate different entities, associated data, and their access control requirements in the framework. In addition, to support the gateway needed for constrained IoT devices such as medical and wearable IoT, we discuss the enhanced ACO architecture with an additional Object Abstraction layer. This framework reflects different authorization needs of and in between different layers and its associated components. Further, in this enhanced architecture we divided the framework into three categories of access control models including object, virtual object and cloud.

In the later part of the chapter, we will focus on the access control requirements in Internet of Vehicles (also known as the vehicular IoT) and connected cars. The prime reason to focus on this domain is its unique characteristic of mobility and dynamic interactions which are different from usually static CPS systems, and requires novel access control solutions. This ecosystem will have multiple cloud and edge systems due to distributed domain and mobile nature of the smart objects. We will discuss the extended access control oriented architecture (E-ACO), which extends to ACO architecture with the introduction of clustered objects. These clustered objects reflect the smart objects which have multiple sensors, similar to smart cars, having 100's sensors inside it having different functionality. Further, it will also reflect possible interactions between sensors in same clustered object or between different object's sensors. Such clustered objects are critical in smart and connected cars ecosystem together with intelligent transportation which has vehicles, traffic infrastructure, road side smart signs and other devices having multiple sensors and actuators in them. This additional layer will reflect the need for controlled communication and data exchange among these entities in the connected ecosystem. Using the proposed architectures, we will also present illustrative examples to highlight the access control needs and issues in cloud and edge assisted IoT and CPS.

2.1.1   Chapter Organization

We will first highlight the fundamental components and reflect distinctive characteristics about different cloud or edge architectures along with various types of objects supported in different IoT and CPS domains. In Sect. 2.3, we will discuss the access control oriented (ACO) architecture designed to capture the needs for access control within and across different layers in the architecture. We will provide two extensions for this architecture in unique domains with gateway/edge in wearable IoT and clustered objects in smart cars. Section 2.4 will illustrate some use cases and how the access control architecture fits into different IoT and CPS systems.

2.2   Primitives for Cloud and Edge Assisted IoT

In this section, we will present the founding principles to support IoT and CPS applications. We will discuss about the different types of smart devices along with cloud and edge architectures are fundamental to support both real time and smart connected ecosystems.

2.2.1   Taxonomy of Smart Devices

We first present a general classification of IoT devices based on three main characteristics—*mobility*: state of their movement, *size* of IoT devices and *nature* of smart devices. Figure 2.1 shows the classification and taxonomy of IoT things based on these three characteristics [1]. First, let us discuss mobility, generally the smart devices have some inbuilt capabilities as well as they inherit the properties of their owners (users who own those devices) or of the entities to which they are attached. Therefore, as per the mobility, we can classify devices as **static** and **mobile**. Static things are fixed and do not have any movement capability, for example, a device cannot move and is restricted to the location of their installation (e.g., a smart camera at home or on a building). Whereas, mobile things are able to move, independently, such as autonomous cars), or dependently with users, such as

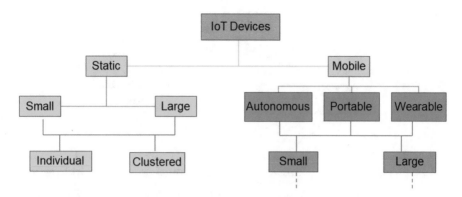

Fig. 2.1 A general classification of IoT devices

wearable smart watches. Thus, they can be further classified into three categories: autonomous which are capable of moving independently, portable which can be carried around, and wearable which can be worn and attached to their owners. Size: IoT devices are of different sizes, from a small tiny sensor to big complex machinery. It is difficult to define definite metrics to categorize IoT things based on the size. However, for simplicity we consider two categories: small and large. For example, any device that can be easily carried by an individual is a small IoT device, such as small sensors or wearable devices. Thus, we consider only small category under portable and wearable. Nature: The third characteristic is the nature of things or devices. The nature of IoT devices depends on their architecture and functionality. Any thing that acts individually to perform a task is an individual IoT device, and a combination of multiple things that operates together to achieve a specific functionality is a clustered IoT device. As the name implies, individual things are made up of a single thing (e.g., a sensor sensing motion), and a clustered device is a combination of small sensors, such as wireless sensor networks works (WSNs) or a smart car that has multiple sensors and actuators. The dashed lines under the small and large devices reflect each of them can further be classified into individual and clustered as in case of the static devices.

Based on various application domains, IoT has started to diverge into different IoT sub-fields, such as Vehicular IoT (VIoT), Medical IoT (MIoT), and Wearable IoT (WIoT). The objective of our IoT device categorization is to provide an overall general classification of heterogeneous IoT devices, and we believe that the above three characteristics are most suitable ones for this purpose. This categorization provides a basis to represent different IoT sub-fields, where distinct nodes in the tree can be combined to realize these sub-fields. For example, VIoT would be a combination of autonomous, large, and clustered IoT devices (sensors and actuators). Similarly, wearable, small, and individual or clustered device categorization can be realized as WIoT, as well as corresponds to MIoT to some extent. Therefore, this classification will enable IoT stakeholders, researchers, and businesses to focus on desired IoT

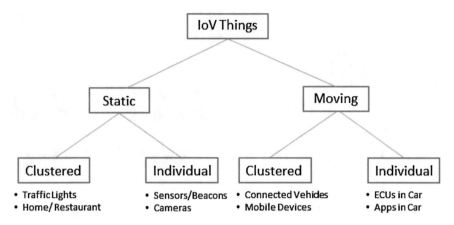

Fig. 2.2 Smart device types with clustered objects

sub-fields and associated security and privacy issues while developing innovative solutions.

Smart ecosystem also envision other kinds of objects (as shown in Fig. 2.2) reflecting their dynamic and mobile behavior [2–5] along with distributed functionality and the applications such ecosystem envisions. For example, in intelligent transportation and connected cars ecosystem also referred to Internet of Vehicles (IoV), there are both static and dynamic objects. Roadside sensors, pedestrian with mobile devices, restaurant and gas station beacons, deer threat warning signs etc. are static in nature and do not move from the position they are installed. In addition, smart cars, connected vehicles, autonomous tractors, etc. are more dynamic and mobile, having their location changing as they move. This means, they can randomly connect and interact with different other devices in their vicinity, depending primarily on their location, and interest. Further, several of the individual objects or smart devices have only one sensor, performing only one service or functionality such as a worker on road sensor. This sensor will only notify if there is *work in progress* on the road ahead. Several smart objects have multiple sensors, like a connected car, or aerial drones in agriculture. They can do multiple functions at the same time, like, a drone can do the thermal imaging at the same time perform weed spray. A smart car has multiple sensors, like TPS (Tire Pressure Monitoring), Radar, blind spot etc. In addition, there are 100's of ECU (Electronic Control Units) for engine, transmission etc. which can also be remotely controlled and commanded. These characteristics are important to distinctively reflect the needs for access control models in such clustered objects as well, and controlled in-clustered object (sensors within the same objects), or across clustered objects (i.e. within vehicles) is important.

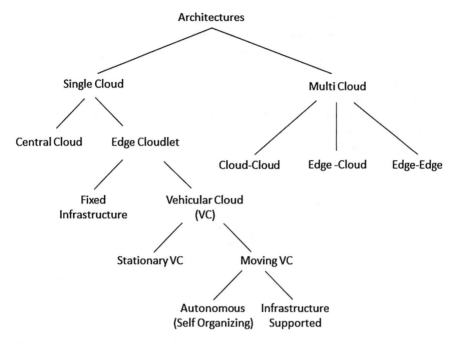

Fig. 2.3 Various cloud and edge architectures for dynamic environments

2.2.2 Cloud and Edge Hybrid Architectures

It is intuitive that with the wide range of applications supported by IoT and CPS, we would need to have both edge and cloud supported systems which will enable real time and near real time applications and services to the end users. Several cloud platforms such as Google Cloud, Amazon Web Services (AWS) or Microsoft Azure will play an important role by offering services and infrastructure by supporting use cases from different smart domains. The use of edge and fog computes is crucial to resolve issues of cloud latency and bandwidth along with communication delays which are not desirable for applications requiring fraction of seconds to respond. In addition, the notion of edge AI will enable local anomaly detection and deployment of trained smart agents, such that the need for communicating with the cloud will be minimal. This also solves the issues of internet connectivity and location sensitivity especially in domains such as smart cars where consistent network cannot be expected in terrains where these smart objects will move. Figure 2.3 illustrate different single and multi-cloud hybrid architectures supported in different IoT and CPS ecosystems. Single cloud architecture requires a central cloud to support different user services, virtual objects, AI and machine learning models etc. and also store data generated by connected devices. However, single cloud systems are not feasible since entities (for example vehicles) may be in wide geographic locations not supported by single central cloud, or may belong to different car manufacturers having their private cloud with which vehicles share data. In such a scenario multi-cloud ecosystem is desirable, where trusted information exchange may take place among different cloud service providers. In addition, edge cloudlets using a fixed infrastructure to support real time applications, or vehicular cloud built in ad-hoc on demand using the storage and compute in vehicles are also possible options, and have been discussed by different researchers in the past. It is also possible to use edge structures for movable objects such as connected cars, and sensors in the vehicles have virtual objects created in the edge, and are transferred as the vehicles move from one place to another. Another notion of vehicular cloud (VC) [6–9] can be created in a shopping mall using the resources of the parked vehicles and offering incentives for the user who participate in such kind of infrastructure. Such resources can also be used to create moving VC where the vehicles can join and leave the cloud as in the nearby region of geographic range. In addition, these moving VCs can be supported by fixed infrastructure (example, a traffic light on the roadside acting as a broker) or moving vehicles in autonomous manner can form a VC. In multi-cloud CPS architectures, we envision to have either multiple clouds, cloud-edge or multiple edge setup. However, we believe single central cloud and multiple edge architectures are a good fit to cover most real time and distributed IoT and CPS applications.

2.3 Access Control Oriented Architectures

In Chap. 1, the general IoT architecture is explained with three main layers: the application layer, one or more middle layers, and the object layer. The division of middle layer is also described, where some earlier proposed IoT architectures point to the middle layer with only one layer like network layer, while others proposed IoT architectures that divide the middle layer to two or more. From the reviewed proposed IoT architecture, we realized that there is a demand for integrating the cloud with the IoT architecture and a benefit to using virtual objects (VO) as a counterpart of physical objects. The access control architecture for IoT that is proposed by Alshehri and Sandhu [10] is described in this section as one of the earliest IoT architecture for access control. This architecture is designed to contribution in proposing access control (AC) models for IoT, and thus this architecture called an AC-oriented (ACO) architecture for the IoT. It emphasizes enabling cloud computing to assist the functionalities of middleware and service management in the middle layer [10].

The ACO architecture for IoT mostly is close to general architecture of the IoT. It is designed to have the two main layers (the application and object layers) that is placed in the most proposed IoT architectures. The middle layer in the general IoT architecture is divided in the ACO architecture into two layers: a virtual object layer and a cloud services layer. Thus, ACO architecture is built with four main layers: an object layer, a virtual object layer, a cloud services layer, and an application layer. Figure 2.4 shows ACO architecture for IoT layers, where application layer

Fig. 2.4 ACO architecture
for the cloud-enabled IoT

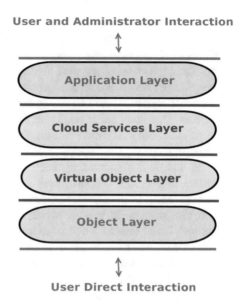

appears in most top and object layer appears at the bottom. The functionalities and components of each layer is presented below.

- **The Object Layer:** The basic functionality of the object layer is analogous to most of the object layers that are reviewed. Object layer accumulate data from surrounding environment to build a wide overview of the data by sharing it with other objects or sending it to the upper level (virtual object layer). There are heterogeneous kinds of objects in this layer including sensors, actuators, cameras, radar that appear individually or form one or more clusters. Objects in this level gather data from physical environment, such as data gathered from temperature sensors, and then send it to upper layer or other objects. In addition to sending data, objects at this layer can receive data (such as control commands from remote user) from upper layer or other objects. A smart light bulb, for example, will not be turned off or change its light color until a coming command direct it to change its state. As a result, objects in this layer can collect data from the environment, changes the state, push other objects to change the state or actuate them, and send it as output or receiving data as input.

  Users can interact with the objects at this layer by pushing the button, or using their mobile phone within the device bluetooth range. As shown in Fig. 2.4, users can interact directly by placing new objects, pressing a button, changing an object place, turning of an object, etc. Moreover, objects at this layer can use communication technologies such as Bluetooth, Wireless, ZigBee, 6LoWPAN, ISA 100, WirelessHart/802.15.4, 18000-7, and LTE [11] to communicate with each other within object layer. Additionally, objects can communicate to their virtual objects (digital counterparts) through the Internet, which is the main idea of IoT technology. However, any direction of communication needs to be authenticated by using authentication technologies such as public key infrastructure (PKI) or digital certificates. Physical objects may receive or send data at any time. Conversely, physical objects could be deliberately or accidentally turned off or on. Thus, the status of physical objects is not known when they are off, and their input/output data is not reached as long as they are disconnected for any reason. However, in the IoT and CPS architectures, it is required to know the status of objects. One solution to reach the status of objects is to have virtual objects of physical objects, which will log the last reported state of the corresponding physical devices.

In addition to supporting reach the last/current status of physical objects, virtual objects can support physical objects in computational tasks and data storage. Most of physical smart devices have constrained computational power and low storage. In another words, physical objects can only achieve basic computational functions and store limited data. However, IoT is based on large sets of collected data, so physical objects require another party to store the vast collected data and perform extensive computational tasks on them. The third party, which is the cloud services layer and the applications supported will be explained later in this section.

- **The Virtual Object Layer:** In the virtual object layer, a persistent current status of objects in the object layer can be presented by virtual objects (digital counterparts) [27] if both are connected. Also, if physical objects are not connected, virtual objects can also present last received status of a physical object, a desired future status, or both the future and last received status. Virtual objects can present all physical objects' services, some of physical objects' services, or one physical objects' service. However, the ACO model assumes that each virtual object represents all services of one connected physical object. Also, the ACO model assumes that only physical objects will have digital counterparts, and there is no digital counterpart for end users (though that may be suitable as the architecture evolve). The virtual objects in this layer provide various solutions to the IoT issues such as scalability, heterogeneity, security and privacy, and identification. Consequently, despite of heterogeneity and locality in the object layer, virtual objects in this layer can communicate with each other. The communication among virtual objects should be controlled for security and privacy issues by using suitable access control mechanisms, such as RBAC [12], ABAC [13–15, 28], or ReBAC [16].

  Physical objects can be associated with virtual objects in multiple methods. The basic method is to associate one virtual object with one physical object (if any) that has one or more services. It results in an associated method called one (or less)-to-one association [17]. However, if on physical object has different services, there could be various virtual objects for different services. It results in an associated method called a one-to-many association. For example, one smartphone could have multiple services and represented by associated to only one virtual object and another way to represent smartphone services is to associated each service to one separate virtual object, e.g. one for touch sensing and one for humidity sensing. Thus, this results in an associated method called a one-to-many association [18]. In addition, another possible association method is to represent a group of physical objects with only one virtual object, where the one virtual object could work as a management or data collection point. A virtual object can manage various physical objects more powerfully with less resource utilization than developing a distributed implementation (many-to-one) [18, 19]. Virtual object can can gather the information of one service from different physical objects (many-to-one) [18]. As a result, some mentioned association methods can be combined to result in an associated method called a many-to-many association [10].

- **The Cloud Services Layer:** This layer is designed to support most of the functionalities associated to the middle layer in the general IoT architecture. The huge amount of data that can be collected with billion smart objects, there is a need to develop means to access, store, and process the huge amount of data collected by these smart objects. Thus, this layer helps in storing and processing the big collected data. The diverse and humongous amounts of collected data in this layer can be utilized intelligently for smart observing and actuation, it can be described to users in more meaningful ways. This presented data can help policymakers (or administrators) to generate new policies or modify existing policies based on the visualized data. Both of collected data and policies are saved in the cloud. Thus, the communication and access between applications and objects are accomplished through the cloud. Furthermore, the cloud services layer also helps in the heavy computational tasks that cannot be handled by the constrained objects. Hence, processes like computation, visualization, and analysis of collected data can be supported by the cloud services layer. Finally, the multi-cloud communication can appear at this layer where information can be shared at a broad level and pursue common goals. Common platform to analyze Big Data is Hadoop where several access control models [20–23, 29] have been proposed to ensure the confidentiality and integrity of the resources .

- **The Application Layer:** The topmost layer in the ACO architecture is the application layer. This layer offers a useful abstract of collected and analyzed information from IoT entities to the users through an interface. Also, users can use the interface to easily communicate with objects and visualize the analyzed information. Administrators also can interact with the IoT system through the interface at application layer to create policies or to update/add policies based on the collected information. Administrators also can manage the communication among objects and virtual objects through the application. Both of users and administrator can communicate with remote physical objects or virtual objects through applications. Users at their work, for example, can turn off a remote light bulb at their home by sending a command to the remote light bulb through the application layer. However, any kind of communication with IoT entities through the application layer should be controlled and authorized by applying suitable access control methods.

2.3.1 Edge Gateway Supported ACO Architecture

Today, cloud computing has become a key enabling infrastructure and technology for IoT. However, there are various application domains in IoT that demand quick response time and low latency. The IoT devices are constrained in nature and significantly low bandwidth. The round-trip time from an IoT device to cloud computing platform may take larger response time that users are not able to use the device efficiently. For an example, a device may be communicating with cloud for completing a certain action, such as publish messages, check authorization policies for making decision on if the action is allowed or denied. In critical application domains, such as healthcare and military field, quick response time and low latency is essential to make decisions in near real-time. Moreover, a reliable connection in such critical domains is also a necessity and intermittent connection with cloud platform may result into costly and catastrophic results. For instance, in the remote field at any location where there is limited connectivity and Internet, a local edge gateway deployed towards the edge of the network can enable local communication, computation, storage, etc. Such edge gateway is small cloud moved towards the edge network. With a rapidly increasing number of IoT devices, edge computing with edge gateway or cloudlets supported architecture is essential for enabling local computation, storage, and analytics towards the edge of the network.

Edge computing is necessary in various IoT domains where low latency and fast response times are a must, such as healthcare where a doctor need to make instant decision based on the IoT data, or military where people can lose lives if the response time is too long for any queries send to the cloud computing platform. Therefore, we need an edge computing supported cloud-enabled IoT architecture. Figure 2.5 shows an enhanced ACO architecture where we introduced the **Object Abstraction** (OA) layer that can provide an abstraction between the Object layer and Virtual Object layer. The edge gateway or cloudlets resides within the OA layer which provides the infrastructure to support local communication and computation.

In the context of wearable IoT, devices are small constrained attached to the user which are continuously collecting their data. The OA layer allows these devices to communicate to an edge gateway device that provides similar services to cloud but at a lower scale. These gateway devices enable to abstract the physical devices' heterogeneity in terms of communication and networking protocols and enables seamless communication between object layer and higher layers in the enhanced ACO architecture. Besides, the OA Layer is added in the architecture to clearly identify different types of physical objects and distinguish between edge and gateway devices and other physical devices—sensors, actuators, etc., in the architecture.

The OA layer is extended from the object layer and is comprised of gateway devices, such as smart phones. It has a unique task to facilitate object to virtual object communication abstracting all the heterogeneity (network and communication protocols) involved in the object layer. We assume as the edge devices become more sophisticated in the near future, the need of an abstraction layer

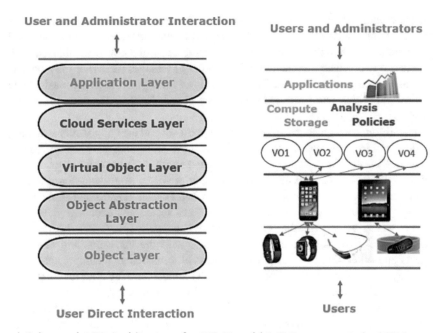

a) Enhanced ACO Architecture for WIoT   b) IoT Components in ACO Layers

Fig. 2.5 An enhanced ACO architecture with edge computing capabilities

may be reevaluated. In the rest of the chapter, the ACO architecture refers to our enhanced ACO architecture, unless otherwise specified. In Fig. 2.5b, various components within each layer and their interactions are shown for a typical wearable IoT scenario with wearable edge devices, gateway devices, virtual objects, cloud services, and applications for monitoring and visualizing the IoT data.

2.3.2 Extended ACO Architecture with Clustered Objects

Mobile IoT smart devices such as vehicular objects and connected cars have multiple heterogeneous devices both individual and clustered together with several on-board applications, which work together to offer services and data driven applications to end users. Several objects have one functionality and are independent of each other, such as the farming sensors [24, 30], traffic lights, beacons road side. Further, clustered objects (COs) have several sensors incorporate in them, for example, an agriculture aerial drone has multiple sensors having different functionality like camera, spraying machine, or a smart vehicle which has electronic control units (ECU), blind spot or tire pressure sensors. The proposed Extended Access Control Architecture (E-ACO) [2] incorporates the distinction of clustered objects to address CPS domains such as manufacturing, transportation or agriculture which supports such *large* objects. Clustered objects are important to understand the communication and data exchange among (across) the clustered objects, such as between two vehicles exchanging Basic Safety Messages (BSM) or intra-objects such as sensors within the car. This distinction is not possible in the basic ACO architecture which was proposed considering physical objects having only single sensor.

In addition, such clustered objects may support several user applications running inside them, for example, a smart vehicle can have safety warning, ice on road or navigation applications, which can communicate with different sensors or devices within the vehicles, or with nearby objects such as with a smart sign-board to alert the driver via seat vibration, or dashboard warning. Further, sensors and applications belonging to one clustered object may be able to access and exchange data from sensors which belong to other clustered objects as well. Figure 2.6 illustrate the extended ACO (E-ACO) architecture together with the corresponding clustered objects and other smart devices in the context of vehicular IoT components in E-ACO layers as shown in Fig. 2.6b. This E-ACO architecture has four layers similar to ACO: Object layer, Virtual Object layer, Cloud services and Application layer, where the communication can happen within a layer (shown as self loop in Fig. 2.6a) and the adjacent layers above and below. Here, we will elaborate on different layers and the additional components introduced in the E-ACO architecture in more details.

- **Clustered Objects in Object Layer:** The object layer introduces clustered objects which have multiple individual sensors or smart objects. The **clustered objects** may also have several built-in applications (like tire-pressure monitoring) installed within them. These applications can communicate with ECUs and sensors in same car (or neighbouring car) to get data and update information to the drivers. Communication can occur between objects (and clustered objects) in the object layer and also with the layers above (virtual object) and below it (user). Communication across objects (within the object layer) among different vehicles or clustered objects is feasible via technologies like dedicated short-range communications (DSRC), Bluetooth, WiFi, and LTE. It is important to note that we have introduce clustered objects as a part of the object layer (not separately) reflecting its binding within the objects, applications and the clustered objects they belong. This relationship among objects and clustered objects is critical, for example, a lane departure sensor in car will have some attributes (like vehicle id) it inherits from the car and such binding is shown by putting them in same layer. Further, these COs have applications associated inside, such as rear vision system in the car which also communicate with sensors and other applications in the system. These applications in object layer of E-ACO is add-on to the object layer in ACO architecture and reflects its importance in CPS ecosystem which is very dependent on in-built applications supported by smart cars and other similar domains.

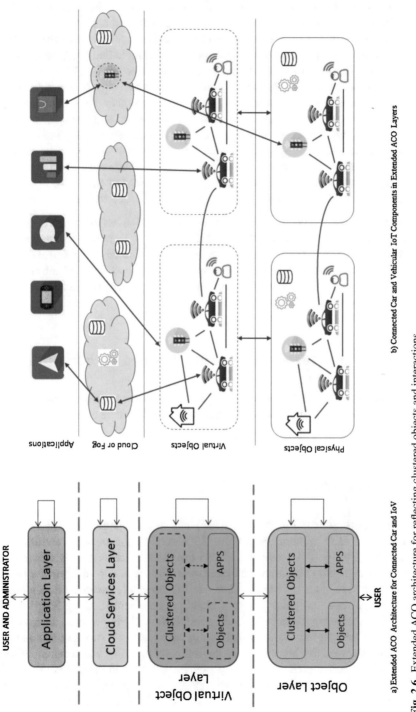

Fig. 2.6 Extended ACO architecture for reflecting clustered objects and interactions

- **Clustered Objects in Virtual Layer:** The virtual layer in E-ACO will have cyber replica of the clustered objects as well as the objects and applications as in the ACO architecture. Physical sensors and objects interact with the cyber replica using secure communications channel and protocols including Message Queuing Telemetry Transport (MQTT), Hypertext Transfer Protocol (HTTP), Constrained Application Protocol (CoAP) etc. As an example, when objects across different clustered objects want to exchange data among each other, the communication will pass through the virtual replicas of these sensors, and then to the other side of the physical objects. Similar communication can be supported for in-built clustered objects applications which can exchange information from objects using their cyber counterparts created in cloud, edge or cloudlets architecture
- **Cloud Services and Application Layer:** The services of these two layers are similar to the ACO architecture discussed in Sect. 2.3, and will not be iterated.

Figure 2.6b reflects how in case of smart transportation having physical objects such as vehicles, cars, traffic infrastructure together with cyber counterparts in virtual objects layer and other E-ACO layers. As can be noted, the physical objects interact with virtual objects and different applications access data and information via cloud or edge services which is pushed by the virtual objects (which can sometimes be created in cloud or local edge also).

2.4 Illustrative IoT and CPS Using Proposed Architectures

In this section, we will reflect two use cases to show how the proposed access control architectures including the *enhanced* and *extended* ACO can be utilized in different IoT and CPS domains.

2.4.1 Remote Patient Monitoring (RPM)

Here, a remote patient monitoring (RPM) use case is presented aligned with the Enhanced-ACO architecture. Figure 2.7 shows all the possible entities and interactions between those entities. In this use case, Alice is a user who has a problem of high blood pressure and uses wearable devices to monitor her body parameters and monitor her health. There are four wearable devices, i.e., a motion sensor, a heart rate and pulse sensor, a blood pressure sensor, and a temperature sensor, that measure specific vital body parameters. All the physical devices are connected to the gateway device which in this case is Alice's smartphone and it is within the OA layer. OA layer provides the first level of access and communication control where user-centric data privacy policies are defined. The smartphone at OA layer can also provide edge computing and analysis capabilities. All these wearable devices have their equivalent virtual objects or digital devices created

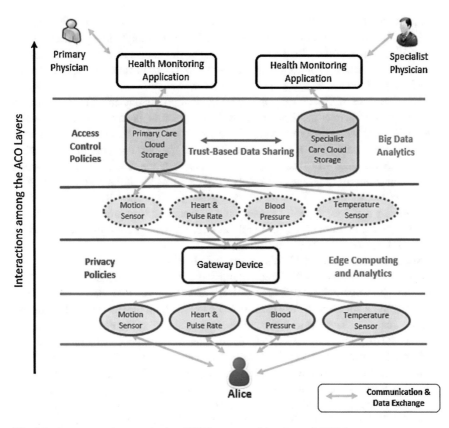

Fig. 2.7 A remote patient monitoring (RPM) usecase with enhanced ACO layers

at the VO layer. There are different types of relationships between virtual object and physical object (e.g., one-to-one, one-to-many, etc.), however, we consider one-to-one association at the VO layer. VO layer facilitates seamless communication between applications and physical devices, and addresses several IoT issues, such as identification, scalability, heterogeneity, security and privacy [25].

These devices collect a huge amount of IoT data and then store and analyze this data in cloud services layer. It is important to understand that OA layer can also allow to store and analyze some portion of data within the edge devices. As shown in the diagram, there are two data storage in the cloud, one for Alice's Primary physician and second for her Specialist physician. All the data by default is stored at her Primary physician's data storage. The Specialist physician would be able to access or request authorization the data securely shared with the Specialist physician only when the need arises especially in a critical situation. Data security and privacy must be maintained based on trust established between the two physicians with the approval and consent of the user. Access control and privacy policies for any access control model, designed for either secure communications or data security and privacy, can be defined at the cloud services layer as a separate authorization service. The cloud services also allows to use Big Data analytics as per the services are available in specific cloud platform. Finally, the analyzed data and results are forwarded to health monitoring applications to show meaningful results to the physicians, which reside at the application layer.

This use case discusses various access control points and interactions between various entities across the Enhanced-ACO architecture layers. These access and interactions between different components need to be secured with dynamic and flexible access control models that can focus on specific components and authorizations associated with those components. For instance, device to device access and permissions of sending and receiving data must be addressed with suitable access control model(s). Thus, an access control (AC) framework is developed for categorizing different types of access, interactions, and communication between different layers of Enhanced-ACO architecture. We will present this framework in detail for CE-IoT in the following chapter.

Another critical aspect of this use case is how would we actually implement such use case in real-world? Thus, authors propose an implementation platform based on [26], using the AWS IoT platform and its services. In this work, authors implemented a smart-home use case in AWS IoT with all the entities and necessary communications. The devices and VOs configurations and authorization policies are defined in cloud. In AWS IoT, *Thing* (the VO in AWS IoT) is created for each device which has a *Thing Shadow* associated with it. The thing provides a set of topics for clients (devices, apps) to publish/subscribe messages. A certificate is created for each device in AWS IoT and is copied onto the physical device. It is necessary to make sure that devices are compliant with the communication protocols of AWS IoT, i.e., MQTT. AWS IoT has its own device gateway that enables secure authentication and bidirectional interactions with edge devices. Moreover, users should be able to define specific privacy-preserving policies, however, this capability is not provided by AWS IoT. There are specific privacy preserving approaches that need to be explored and applied as per the use case scenario. We can also store the IoT data generated in a DynamoDB database in cloud so that any desired analysis can be performed using cloud analytic services. The applications that physicians use can also be deployed in cloud.

2.4.2 *Intelligent Transportation System (ITS)*

Vehicular IoT and Intelligent Transportation [2, 4, 5, 28] is a novel domain where networking and communication among vehicles, cars, traffic infrastructure, homes, restaurants, pedestrians or eventually everything is proposed. The main objective of Internet of Vehicles (IoV) is inter-connectivity and interaction among smart objects in which vehicles and smart ITS infrastructure are most important. The vision of this connected CPS is to offer safety, convenience and value added services to the end users. As stated by the Wikipedia, *A Connected car (or Connected Vehicle*

(CV)) is a car equipped with internet access and usually also with the wireless area network. This allows the car to share internet access with other devices both inside as well as outside the vehicle. The interactions among vehicles, cars and traffic infrastructure, automatic braking and emergency calling, driving assistance and autonomous driving, parking areas, weather and accident warnings, E-toll, and predictive maintenance, are some of the most critical and desirable applications in today's connected vehicles and ITS. These smart objects have over 100 ECUs and more than 100 millions lines of code in support of such smart services. These connected vehicles (CVs) have controller area network (CAN) bus, FlexRay, Ethernet and other protocols which are used for ECU communication within the vehicle among applications and different on-board sensors. Broadcast method is used to send messages to different ECUs attached via the CAN bus. Several different buses connect each other via a TCU (Telematics Control Unit) which acts as a gateway and offers an interface to external environment. These connected vehicles produce large volumes of data which is exchanged within and across different vehicles, tempting community to refer these CVs as *smartphones on wheels.*

As shown in Fig. 2.6b, the ITS ecosystem fits perfectly into the E-ACO architecture. At the **Object layer**, vehicle and smart cars represent clustered objects, which communicate with the single objects (sensors) and application in the vehicles. One example of such interaction in object layer is BMWs' *connect* application in phone which reads address from phone and send to the car navigation system, or Vehicle-to-Vehicle (V2V) Basic Safety Message (BSM) which is exchanged using the Dedicated Short Range Communication (DSRC) protocol. Several other applications including cabin monitoring system, safety alerts, infotainment systems, tire-pressure monitoring which communicates with the sensors in the tire, etc are applications which are supported by connected vehicles and can also capture data from in-vehicle or nearby sensors to support services. These on-board applications in the E-ACOs object layer is an addition to the object layer is base ACO architecture, and highlights its importance in ITS ecosystem which is interaction among different entities supported by the smart vehicles. In the **Virtual Object layer** ITS creates cyber replicas of physical entities including both clustered and single objects. The most important smart entity in ITS, a smart vehicle, is usually in motion which passes through different terrains and geographical areas with unstable and intermittent internet connectivity all times. In such cases, it is critical to create cyber entity of smart objects such as cars and infrastructure components in cloud or edge infrastructure, which can report the last state of the objects along with the desired state information of different sensors, which can then be updated to physical objects counter part once vehicle gets connectivity to the internet again. As an example, if some maintenance issue is diagnosed in the power train control ECU of the vehicle, and the mechanic wants to send to the sensor to control the air-fuel ratio. In such a scenario, if the vehicle has connectivity, the mechanic can directly interact with the ECU and updates it state immediately. However, this may not be the case when internet is unavailable and the latest state of the ECU is not available. In this case, the mechanic will send the message to the virtual entity of ECU created in the cloud, which will then send this updated desired state to the physical ECU when vehicle gets connected and synced to the virtual ECU. We envision the virtual object layer in E-ACO architecture will have one or many cyber entity (virtual object or device shadow) for both clustered and individual objects, such as one supported by AWS named shadows. This also limits what data different entities can access from various on-board sensors.

Cloud and Application layer offers storage and processing of data for different applications and services for the end users. In addition, Over the Air (OTA) updates for firmware and other software components provided by different manufacturers of the vehicles are supported through the cloud services layer where only authorized users, manufacturers and third party applications are allowed to issue OTA. User and various applications can access the data pushed from the sensors into the cloud by smart infrastructures and vehicles for offering value added services to customers. The proposed E-ACO architecture assumes to have both central cloud and fog (instantiated by vehicular cloud) component in ITS ecosystem but are collectively represented as cloud services.

As reflected by the description of the ITS use case and how it fits in the E-ACO architecture, each of the supported layers have corresponding entities in ITS. In the next chapter, we will show the need for access control based on this ITS use case and the communication it requires among different entities within and across layers using the authorization framework.

2.5 Summary

In this chapter, we discussed access control focused IoT architecture, called ACO, which is divided into four layers: the object layer, the virtual object layer, the cloud layer, and the application layer. The chapter first provide the taxonomy of different edge and cloud assisted architectures along with different types of smart objects supported. We then extended the ACO architecture, to support both edge and multi-cloud assisted CPS architectures. We proposed two extensions, including the cloud enabled edge gateway supported architecture for applications such as wearable IoT and an extended ACO (E-ACO) architecture which introduces the novel concept of clustered objects (cars, infrastructure, home), which have several individual smart objects, sensors and applications. The architectures discussed in this chapter will be used as a reference to build access control models for cloud and edge supported Internet of Things and Cyber Physical Systems in the following chapters.

References

1. Bhatt, S., Patwa, F., & Sandhu, R. (2017). An access control framework for cloud-enabled wearable internet of things. In *2017 IEEE 3rd International Conference on Collaboration and Internet Computing (CIC)* (pp. 328–338). IEEE.
2. Gupta, M., & Sandhu, R. (2018). Authorization framework for secure cloud assisted connected cars and vehicular internet of things. In *Proceedings of the 23nd ACM on Symposium on Access Control Models and Technologies* (pp. 193–204).

3. Gupta, M. (2018). *Secure Cloud Assisted Smart Cars and Big Data: Access Control Models and Implementation.* Ph.D. thesis, University of Texas at San Antonio.
4. Gupta, M., Benson, J., Patwa, F., & Sandhu, R. (2019). Dynamic groups and attribute-based access control for next-generation smart cars. In *Proceedings of the Ninth ACM Conference on Data and Application Security and Privacy* (pp. 61–72).
5. Gupta, M., Benson, J., Patwa, F., & Sandhu, R. (2020). Secure V2V and V2I communication in intelligent transportation using cloudlets. *IEEE Transactions on Services Computing.* https://doi.org/10.1109/TSC.2020.3025993.
6. Eltoweissy, M., Olariu, S., & Younis, M. (2010). Towards autonomous vehicular clouds. In *International Conference on Ad hoc Networks* (pp. 1–16). Springer.
7. Gerla, M. (2012). Vehicular cloud computing. In *2012 The 11th Annual Mediterranean Ad hoc Networking Workshop (Med-Hoc-Net)* (pp. 152–155). IEEE.
8. Gerla, M., Lee, E.-K., Pau, G., & Lee, U. (2014). Internet of vehicles: From intelligent grid to autonomous cars and vehicular clouds. In *2014 IEEE World Forum on Internet of Things (WF-IoT)* (pp. 241–246). IEEE.
9. Olariu, S., Khalil, I., & Abuelela, M. (2011). Taking vanet to the clouds. *International Journal of Pervasive Computing and Communications.* https://doi.org/10.1145/1971519.1971522
10. Alshehri, A., & Sandhu, R. (2016). Access control models for cloud-enabled internet of things: A proposed architecture and research agenda. In *2016 IEEE 2nd International Conference on Collaboration and Internet Computing (CIC)* (pp. 530–538). IEEE.
11. Al-Fuqaha, A., Guizani, M., Mohammadi, M., Aledhari, M., & Ayyash, M. (2015). Internet of things: A survey on enabling technologies, protocols, and applications. *IEEE Communications Surveys & Tutorials, 17*(4), 2347–2376.
12. Sandhu, R. S. (1998). Role-based access control. In *Advances in Computers* (vol. 46, pp. 237–286). Elsevier.
13. Jin, X., Krishnan, R., & Sandhu, R. (2012). A unified attribute-based access control model covering dac, mac and rbac. In *IFIP Annual Conference on Data and Applications Security and Privacy* (pp. 41–55). Springer.
14. Gupta, M., & Sandhu, R. (2016). The GURA$_G$ administrative model for user and group attribute assignment. In *International Conference on Network and System Security* (pp. 318–332). Springer.
15. Gupta, M., & Sandhu, R. (2021). Reachability analysis for attributes in abac with group hierarchy. Preprint. arXiv:2101.03736.
16. Cheng, Y., Park, J., & Sandhu, R. (2012). Relationship-based access control for online social networks: Beyond user-to-user relationships. In *2012 International Conference on Privacy, Security, Risk and Trust and 2012 International Confernece on Social Computing* (pp. 646–655). IEEE.
17. Langheinrich, M., Mattern, F., Römer, K., & Vogt, H. (2000). First steps towards an event-based infrastructure for smart things. In *Ubiquitous Computing Workshop (PACT 2000)* (p. 34).
18. Nitti, M., Pilloni, V., Colistra, G., & Atzori, L. (2015). The virtual object as a major element of the internet of things: A survey. *IEEE Communications Surveys & Tutorials, 18*(2), 1228–1240.
19. Römer, K., Schoch, T., Mattern, F., & Dübendorfer, T. (2004). Smart identification frameworks for ubiquitous computing applications. *Wireless Networks, 10*(6), 689–700.
20. Gupta, M., Patwa, F., Benson, J., & Sandhu, R. (2017). Multi-layer authorization framework for a representative hadoop ecosystem deployment. In *Proceedings of the 22nd ACM on Symposium on Access Control Models and Technologies* (pp. 183–190). ACM.
21. Gupta, M., Patwa, F., & Sandhu, R. (2017). Object-tagged RBAC model for the hadoop ecosystem. In *31st Annual IFIP WG 11.3 Conference on Data and Applications Security and Privacy (DBSec)* (vol. 10359, pp. 63–81). Springer Lecture Notes in Computer Science.
22. Gupta, M., Patwa, F., & Sandhu, R. (2018). An attribute-based access control model for secure big data processing in hadoop ecosystem. In *Proceedings of the Third ACM Workshop on Attribute-Based Access Control* (pp. 13–24).
23. Awaysheh, F. M., Alazab, M., Gupta, M., Pena, T. F., & Cabaleiro, J. C. (2020). Next-generation big data federation access control: A reference model. *Future Generation Computer Systems, 108*, 726–741.

24. Gupta, M., Abdelsalam, M., Khorsandroo, S., & Mittal, S. (2020). Security and privacy in smart farming: Challenges and opportunities. *IEEE Access, 8*, 34564–34584.
25. Alshehri, A., & Sandhu, R. (2017). Access control models for virtual object communication in cloud-enabled IoT. In *2017 IEEE International Conference on Information Reuse and Integration (IRI)* (pp. 16–25). IEEE.
26. Bhatt, S., Patwa, F., & Sandhu, R. (2017). Access control model for AWS internet of things. In *International Conference on Network and System Security* (pp. 721–736). Springer.
27. Cathey, G., Benson, J., Gupta, M., & Sandhu, R. (2021). Edge centric secure data sharing with digital twins in smart ecosystems. *Preprint arXiv:2110.04691.*
28. Gupta, M., Awaysheh, F. M., Benson, J., Alazab, M., Patwa, F., & Sandhu, R. (2020). An attribute-based access control for cloud enabled industrial smart vehicles. *IEEE Transactions on Industrial Informatics, 17*(6), 4288–4297.
29. Gupta, M., Patwa, F., & Sandhu, R. (2017). POSTER: Access control model for the Hadoop ecosystem. In *Proceedings of the 22nd ACM on Symposium on Access Control Models and Technologies* (pp. 125–127).
30. Gupta, M., & Sandhu, R. (2021). Towards activity-centric access control for smart collaborative ecosystems. In P*roceedings of the 26th ACM Symposium on Access Control Models and Technologies* (pp. 155–164).

Chapter 3
Authorization Frameworks for Smart and Connected Ecosystems

3.1 Introduction

Security and privacy in IoT and cyber-physical systems are key factors that will foster and support its wider adoption at the consumer level and adapt into different smart domains. Access control has been the key technological mechanism to controlling access and communication among subjects and objects in the system. The dynamic and distributed nature of certain IoT domains including intelligent transportation and connected vehicles brings in challenges to secure the CPS ecosystem. Broad attack surface and numerous external interfaces along with the intrinsic characteristics of IoT makes it even harder to ensure security and privacy of the components and data inside. Access controls are important to restrict unauthorized access to data, sensors, applications, infrastructure and other resources in domains such connected cars and wearable IoT. To understand, identify and develop a comprehensive set of access control models, we must develop authorization frameworks that capture different types of communications and data exchange within the different layers of ACO and its extension architectures. The layers of the ACO architecture encapsulate various entities, such as users, edge objects, gateway objects, virtual objects, cloud services, applications, and administrators, and these entities further comprise of other sub-entities. Further in connected cars and vehicular IoT systems, applications for example, [21] and [22] support vehicles (and its sensors) to access data from not only its own sensors but also vehicle sensors in the vicinity. The Basic Safety Messages (BSMs) exchanged among vehicles and smart objects, must be trusted and verified. In addition, clustered object (in-vehicle) interaction with the CAN buses between sensors and ECUs within the cars and applications should be secured. This exchange of messages must be authorized to ensure confidentiality and integrity of vehicle's and user's personal data, and to prevent remote (or physical) control of connected smart entities. A single access control model would not be sufficient to capture all the access control requirements of different layers (and their associated entities) in the ACO architecture. In this

M. Gupta et al., *Access Control Models and Architectures For IoT and Cyber Physical Systems*, https://doi.org/10.1007/978-3-030-81089-4_3

chapter, we define an access control framework that reflects authorization needs at various layers of extended and enhanced ACO architecture discussed earlier. This will also enable us to discuss some access control models and authorization approaches relevant for wearable and vehicular IoT ecosystem which have different needs and access control requirements.

3.1.1 Chapter Organization

We will first discuss the access control framework for wearable IoT discussing about the various interactions based on the object abstractions and different layers discussed in enhanced ACO architecture. This section will also elaborate on the access control models at different layers based on the framework. In Sect. 3.2 we will also discuss a remote monitoring use case and the need for different access control models. In Sect. 3.3 authorization framework with respect to moving smart objects and clustered objects is discussed along with single and multi-cloud systems. Section 3.4 highlights the objectives of the framework enumerating the research agenda and the need for different fine grained access control solutions.

3.2 Access Control Framework for Cloud Enabled Wearable IoT

Internet of Things (IoT) comprises of various sub-domains. Wearable IoT (WIoT) is a sub-domain of IoT which comprises wearable IoT devices and users who wear these devices. Wearable devices are tightly associated with the users who use these devices in their daily lives. Security and privacy are primary concerns for WIoT users since wearable devices collect a huge amount of user data including sensitive user data, such as user health data, location, etc. Therefore, it is essential to address security and privacy issues in WIoT to enable its wide adoption and continued success in the future. Access control models and mechanism are key technologies to secure WIoT devices and applications. In this section, we primarily focus on access control and authorization in WIoT. Wearable devices are generally small and constrained devices that utilize a cloud-enabled IoT (CE-IoT) architecture as discussed in Chap. 2. We adapt CE-IoT architecture from a WIoT context and name it as the Cloud-Enabled Wearable Internet of Things (CE-WIoT). In the previous chapter under Sect. 2.3, we presented the five-layered Enhanced ACO architecture with *Object (O) Layer, Object Abstraction (OA) Layer, Virtual Object (VO) Layer, Cloud Services (CS) Layer,* and *Applications Layer.* We will use these layers and the communication among different entities to propose access control framework, and set forth our agenda for future chapters.

Fig. 3.1 Interactions of
entities between enhanced
ACO layers

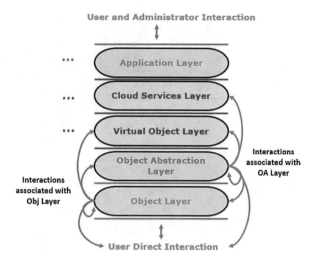

3.2.1   Access Control Framework

Figure 3.1 shows the five layers of the enhanced ACO architecture and various interactions, which include access control interactions and data communications, within each layer and between different adjacent and distant layers of the architecture. The new OA layer is critical in the context of IoT domains such as WIoT as it introduces the edge computing capabilities in the architecture, which offers real time decision capability for resources crunched IoT. Edge gateway devices reside at this layer and enable local computation and communications between wearable devices which have lower bandwidth and also demand low latency and fast response time. However, in the multi-layered architecture, there are various interactions (access and data communication) between these layers which require security by developing appropriate access control and communication control models for specific interactions.

To develop a comprehensive set of access and communication control models for CE-WIoT, we need an access control (AC) framework that can illustrate and categorize different types of access requests and data communications within and among the five layers of the enhanced ACO architecture. Each layer incorporates various entities, such as users, physical smart objects, edge gateway objects, virtual objects, cloud services, applications, and administrators, and these entities may further comprise other sub-entities. It is very difficult to address all the access control requirements of CE-WIoT architecture and its various layers with a single access control model. Therefore, we develop an framework for controlling access and data communications between various entities in the CE-WIoT architecture. Previously, many access control models [1–10] have been proposed for IoT and CPS in the academic literature. Ouaddah et al. [11] provides a comprehensive survey of IoT access control models. The diverse and dynamic nature of IoT requires a unified

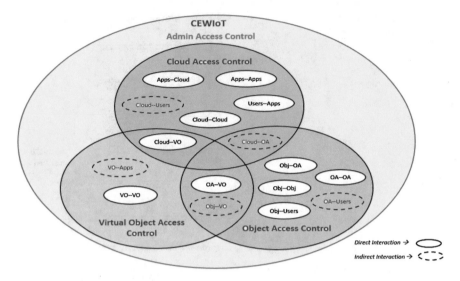

Fig. 3.2 Access control framework based on interactions in the enhanced ACO architecture

access control framework for grouping different types of IoT access control models focusing on distinct IoT components and their specific access control requirements.

Figure 3.1 depicts different types of possible interactions associated with Object and OA layer explicitly, and interactions of other layers follow the same pattern which are represented as dots in the figure. In this architecture, we consider that each layer can interact with itself and its adjacent layers upto two levels both above it and below it. For example, the Object layer interactions are: (i) with itself (Obj–Obj), (ii) with users (Obj–Users), (iii) with OA layer (Obj–OA), and (iv) with VO layer (Obj–VO). There are many of these interactions associated with each layer where each one of them represents access control and authorization points in CE-WIoT. To facilitate the development of access control models that can address specific access and communication control requirements of each layer, three categories of access control models are created: (i) Object Access Control, (ii) Virtual Object Access Control, and (iii) Cloud Access Control, which comprises our Access Control (AC) framework for CE-WIoT.

Figure 3.2 shows the AC framework encompassing all the possible interactions in enhanced ACO architecture. In the enhanced ACO architecture, we identify two modes of interaction between any two layers: *direct interaction* (DI) and *indirect interaction* (IdI). DI represents interaction within this layer and between its immediate adjacent layers; and IdI represents interaction with second level of adjacent layers above and below that layer. In the Fig. 3.2, DI are shown as solid ovals and IdI are shown as dashed ovals. There are some intersections between each of the access control categories which includes the interactions between any two category of models, such as *OA–VO*, and *Obj–VO* which belong to both Object AC and Virtual Object AC models. Thus, we have overlap between three AC categories in the framework.

The outer most circle incorporates the administrative access control which represents admin access within the entire CE-WIoT space, and administrative access control models can be designed for each one of the three AC categories and their respective interactions. The access and communication between layers of the enhanced ACO architecture is controlled by operational access control models. Researchers can focus on specific interactions (one or more) and develop specialized access and communication control models for those interactions. In the following subsections, we present the AC categories of the AC framework for cloud-enabled WIoT.

3.2.1.1   Object Access Control Models

The object access control models encapsulate access and authorizations at two layers—the Object layer and the Object Abstraction (OA) layer and their adjacent layers upto two levels in the layered CE-WIoT architecture. The physical IoT devices are within the Object layer, and gateway IoT devices, which have more resources and capability compared to small physical devices in object layer devices and can enable local computation, storage, analytics, and communication between physical smart objects, are in the OA layer models. Access control models that focus on access and communications, such as data access and transfer within and outside of these layers are grouped into the object access control models category. It encapsulates the following interactions in this category: Obj–Obj, Users–Obj, Obj–OA, Obj–VO, OA–OA, User–OA, OA–VO, and Cloud–OA.

3.2.1.2   Virtual Object Access Control Models

The access control models developed for virtual objects (VO) access and communication control between VOs, such as VO-to-VO, and for interactions between VO and other layers can be grouped into the Virtual Object AC models. These models mainly focus on interactions to and from the VOs, and include three direct interactions—VO–VO, OA–VO, Cloud–VO, and two indirect interactions—VO–Apps and Obj–VO. Alshehri and Sandhu [13] have proposed access control models for virtual object communication, which will be discussed in Chap. 5.

3.2.1.3   Cloud Access Control Models

The cloud services layer allows IoT to leverage its practically unlimited storage, computation, and analysis capabilities. It provides the flexibility and scalability needed for IoT [12]. The cloud is capable of hosting many IoT components. For example, AWS IoT hosts a device gateway, virtual objects, cloud services, and cloud applications. Thus, the access control models in this layer are more complex and may significantly overlap with above two categories. The interactions which need to be secured here are: Cloud–VO, Cloud–OA, Apps–Cloud, Users–Cloud, Cloud-to-Cloud, Users-to-Apps, and Apps–Apps. We include the applications layer interaction within this category of models, since applications mainly utilize the data stored and analyzed in the cloud to provide IoT services to the users. Also, these applications are often cloud applications and database servers hosted in the cloud. Any access control model developed for CE-WIoT can be easily mapped to one of the above three AC categories and may address authorization related to all the interactions (small circles inside a category) or a subset of the interactions relevant to that category. Our AC framework can be easily adapted for a general CE-IoT architecture considering the original ACO architecture and relevant interactions.

Appropriate access control models and architectures are needed to fit into the different interactions in AC framework. Broadly, these access control models can be categorized in *Operational* and *Administrative* models, as also discussed in Fig. 1.3 in Chap. 1. The operational models offer secure access and control of resources and different critical objects in an information system. These models also ensure secure data access. On the other hand, administrative models are primarily used by the security or system administrators to manage the underlying primitives and parameters that are used in operational model. Further, in several systems only administrators have the authority to define and update access control security policies used in operational model. With respect to our framework, the three categories will include both operational and administrative models. The established role based, attribute based models and extensions can be used as both operational and administrative model for different applications and services. We will discuss more of these models in the subsequent sections of this chapter.

3.2.2   RPM Wearable IoT Use Case

Figure 3.3 shows the sequential representation of a remote health monitoring use case. This provides a more clear picture of the data transfer between different entities with respect to specific actions. As shown in the Fig. 2.7 (discussed in Chap. 2), Alice uses four wearable devices—a motion sensor, a heart rate and pulse sensor, a blood pressure sensor, and a temperature sensor, and these devices authenticate and connect to a gateway device. The devices send data to the gateway, which then sends the collected data to corresponding VOs that have been created in the *Primary* physician's cloud. All the data being sent from devices is stored in the database in cloud and can be analyzed. Based on the analytics results, health monitoring apps provide meaningful insights to the primary physician. In a normal case, primary physician sends commands and recommendations to Alice by sending messages to her devices through the VOs. If there is an emergency, then ambulance services are notified and immediate medical help is sent to Alice. Moreover, the *Specialist* physician would be informed if there is a need based on some predefined conditions

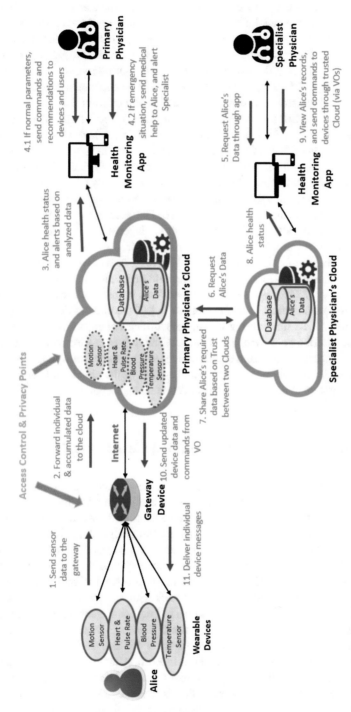

Fig. 3.3 A sequential diagram of interaction of remote patient monitoring example

in the system. Alice's health results are shared with the specialist physician's cloud as needed given there predefined trust established and appropriate access control policies are defined. The specialist physician can also send commands to the devices and recommend Alice to schedule a visit. The gateway device is responsible to make sure that messages being delivered accurately from both directions. In real-world hospital, the setup and configuration devices would be dependent on the user, i.e., Alice, and the administrators, such as cloud and health care admins.

As we discussed different components at different layers of the enhanced ACO architecture, the interactions within and among different layers also need to be secured. In the above scenario, the edge wearable devices associated with Alice should only be allowed to communicate with a gateway device that is owned by Alice or that Alice trusts fully. At the same time, the gateway device also needs to authenticate the edge wearable devices and allow authorized communication and data exchange with respective virtual objects. The access control models which would address such authorizations at the Object layer and OA layer and among their adjacent layers are Object AC models. In our use case, device to device communications are not being considered. However, in real world scenarios (e.g., wireless sensor networks) device to device communication are possible without having a gateway device. For securing interactions within each layer, appropriate access control models need to be developed. In [2], the authors developed several access control models for securing virtual object to virtual object interactions using the publish/subscribe paradigm. The cloud AC models comprise models designed for controlling access to and from cloud services and resources, as well as any access control model developed for securing data [24, 25] in the cloud, and for enabling secure collaboration and data sharing between tenants, accounts, or clouds.

3.3 Framework for Smart Connected Cars Ecosystem

Intelligent Transportation and smart cars have some characteristics similar to other IoT and CPS domains including data communication, device interaction, cloud and edge assisted architectures and secure data sharing. However, as illustrated in Fig. 3.4 several characteristics such as dynamic and swift change in the interaction among smart entities, prompt topological structure change, mobility of smart objects (aka vehicles), geographical scale of the distributed system etc. are certain explicit features which offer distinctive characteristics from other IoT and CPS domains, which results in different and new requirements with respect to cybersecurity and privacy. In addition, most of the applications in intelligent transportation and smart cars are very location and time centric, for instance, if there is an accident on road in your car vicinity, you must get alerts; or if there is deer threat warning or restaurant sending coupons to your car dashboard. Another set of examples such as basic safety messages about congested roads or ice of the road warnings etc. are some applications which are imperative in this dynamic and mobile CPS domain, making its access control and security requirements very unique. In Chap. 2, we also

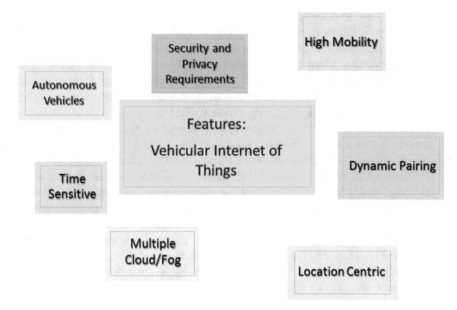

Fig. 3.4 Next generation smart cars distinguishing characteristics

discussed different kinds of objects supported by this domain including the novel notion of clustered objects (such as a smart car), which have objects (sensors and ECUs along with smart applications) within it. These distinctive features are critical to identify and understand since it enables to define access control framework and eventually the family of access control models.

There are several applications and services which are envisioned and supported by smart city and intelligent transportation initiative such as the following:

- *Advertisement Alerts, Information and Entertainment:* Insurance companies have extensively started using driver's behavior and skills based vehicle insurance premiums. They capture data from the sensitivity of the mobile devices, movement of time, and other factors. Soon, with fully integrated next generation smart cars, the same capability can be offered without mobile phones, offering more accurate and customized services to the users. In addition, real time parking services offer just in time and update information about the availability of parking spots in a garage, and that information can be communicated with the smart vehicle, to take it to the exact spot. Restaurants, eateries and even gas stations can send notifications and alerts from the nearby smart beacons to offer coupons and advertisement services. In addition, car-pooling and connected driving applications, live weather and safe routes are some other services next generation intelligent transportation has envisioned in the near future.
- *Accidental Safety and Road-side Assistance:* The communication among different entities including the vehicle to vehicle (V2V), vehicle to infrastructure

(V2I), vehicle to pedestrian (V2P) and eventually vehicle to everything (V2X), several applications offer real time service and information about different smart objects and traffic to control lane departure, vehicle speed, in-lane position control, deer threat warning, or road work ahead warnings from the nearby traffic infrastructure and signboards. In addition, information and alerts about inclement weather in non-ideal driving conditions, assistance and route information in case of high impact and low frequency events (HILF) such as tornadoes, shootings or other critical events can utilize ITS infrastructure for timely exchange of messages and alerts in the system. Further, blind spots, or pedestrian needing assistance can use mobile phones to exchange safety messages including their location, speed and direction with incoming nearby smart vehicles such as while crossing roads, to ensure safety of the pedestrian as well as humans in the vehicles. These messages and information exchange will become more critical with fully autonomous vehicles rolled out in near future.

- *Remote Diagnostic and Vehicle Maintenance:* Health of the vehicle and subsystems will be a key concern in autonomous vehicles, and to maximize road time. This will require for real time monitoring of the vehicles for any safety issues and remote diagnostics mandatory for such vehicles. Vehicle manufacturers and diagnostic service providers can check current health status, the meaning of trouble codes, the potential severity of underlying issues, and actions that can be taken. This will enable a broad range of vehicle health indicators, including fuel level, tire pressure, oil temperature, engine temperature, RPM, and several others. In addition, these services help in fleet-management by reducing the cost of operations through preventive vehicle maintenance, remote vehicle diagnosis and support from original equipment manufacturer service team, reduction in fuel consumption, analyzing and correcting driver behavior. Once the vehicle is on-road every vehicle acts as a test vehicle—any vehicle can be monitored for any parameter. This can help in product planning and design, improving the vehicle performance and reliability. Vehicle sensor data can be send to cloud for processing to predict vehicle mechanical issues. Over the air (OTA) updates can also be issued by manufacturer for fixing car firmware which will obviate the need to go to mechanic. Fleet management applications provide real-time telematics, driver fatigue detection and package tracking.

3.3.1 Access Control Framework

The applications and services offered in the mobile and dynamic ITS ecosystem is due to the interaction among several moving components, making it hard to understand the diverse access control requirements along with different decision and enforcement points due to cloud and edge assisted architectures. The extended ACO architecture discussed in Chap. 2 highlights the various layers and possible communication scenarios among entities in 4 different layers and within the

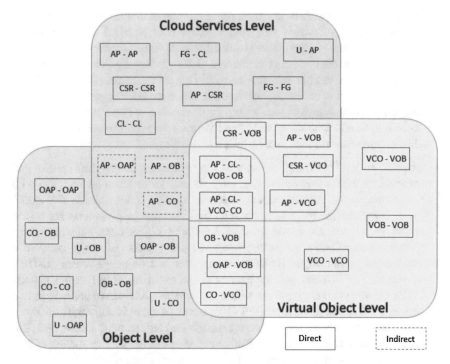

U: User **CO:** Clustered Objects **OB:** Objects **OAP:** Object Layer Applications **CL:** Cloud **FG:** Fog
CSR: Cloud Services **VCO:** Virtual Clustered Objects **VOB:** Virtual Objects **AP:** User Applications

Fig. 3.5 Access control points in cloud and edge assisted smart cars

same layer also. Based on this architecture, we have primarily categorized the
communication scenarios into three categories: *Object* Level, *Virtual Object* Level
and *Cloud Services* Level, as illustrated in Fig. 3.5. Since several value added
services are cloud based applications which harness the resources in cloud, we
have brought together the interactions among different entities within cloud and
applications into single cloud services level category. As elaborated in the E-ACO
architecture, entities in each layer communicate within themselves (i.e. the other
entities in the layer) and also with the components of layers which are adjacent to
it. This results in two different types of interactions, *direct* and *indirect*, as marked
with the solid and dashed boxes in the authorization framework in Fig. 3.5. The
interaction among entities at the adjacent layers (above and below) is considered
direct communication, whereas interaction among entities two or more layers above
or below in E-ACO is considered as indirect communication. As an example,
communication between different smart vehicles (which are clustered objects), and
with the sensors and ECUs within the car (considered as objects inside the clustered
objects) is termed as direct, since they belong to the same physical objects layer.
On the other hand, interaction between remote applications via cloud in application

layer and different sensors or smart object at the physical layer will be considered as indirect since applications will communicate with physical objects via their corresponding virtual entities created in cloud or local edge infrastructure. It must be noted that it is possible to have several interactions overlapping in two categories of the framework, for example, interactions within the cloud service (CSR) and virtual object (VOB) is part of both cloud services and virtual object category. Hence, in the figure, you will notice several interactions corresponding to two or categories of the authorization framework. We have defined the authorization framework categories and ITS communication scenarios as follows.

- **Object Layer:** This category captures the interaction of physical entities and applications within themselves (i.e. other objects at the same layer) along with the entities in the adjacent layers which in E-ACO architecture is the virtual object layer and the end users, who can directly issue or press a sensor button in the smart car. As shown in Fig. 3.5 under object level category, some of these interactions are between two clustered objects (CO-CO), between a user and clustered object (U-CO), among user and on-board sensor (U-OB), among various sensors and applications running in the smart vehicles (OB-OAP), between two applications within the vehicle (OAP-OAP) and within two sensors or ECUs in the vehicle (OB-OB). Based on these identified interactions, it is clear that different access control models and enforcement mechanisms are needed to authorize this communication and data exchange. This includes CAN bus communication to be controlled to enable authorization with the ECUs and sensors in the vehicle, as well as the basic safety messages (BSMs) which are exchanged among vehicles using the Dedicated Short Range Communication (DSRC) protocol. This is critical to ensure the integrity and authenticity of the messages as well as to maintain confidentiality of information shared among different moving entities on the road.

- **Virtual Object Layer:** This category captures direct communication and inter-actions among different virtual objects and entities, with the cloud services along with the end user applications via the cloud. This includes, for example, communication between different virtual clustered objects (VCO-VCO) such as smart cars virtual objects interaction which is usually through the central cloud to which both the vehicles are associated and maintain their cyber replicas. This also includes interaction between virtual objects (VOB-VOB) like the virtual sensors and ECU which can be created at the edge within the vehicle or at a central cloud, between end user applications and virtual objects (AP-VOB), among cloud services and different clustered (or simple) virtual objects (CSR-VOC, CSR-VOB), or between the virtual clustered object (VCO) and the sensor of the nearby vehicle (VCO-VOB) in the cloud to support applications which require data from nearby vehicles. In addition, the interactions in this layer overlap some of the interactions with the object layer since it involves components from the physical layer also. Such interactions include between CO and its virtual CO (CO-VCO), between applications in the vehicle and virtual sensors in the cloud (OAP-VOB) etc. Most of these interactions are controlled with the publish subscribe topic based protocol such as MQTT or communication technologies such as DDS[1] or through HTTP[2], CoAP[3]. Alsehri and Sandhu [2, 13] proposed family of access control models for virtual objects communication using the topic based MQTT protocol applying and extending CapBAC (Capability based access control), ACLs and ABAC.

- **Cloud Services Layer:** Cloud offers *infinite* capabilities providing necessary processing, and storage of humongous amounts of data in a variety of formats, and to enable them to be used by various end user services and applications which harness the true potential of the connected vehicular and smart CPS ecosystem. In addition, most of these applications are also cloud based with their software and hardware components are supported by the cloud service providers. Therefore, we have bundled together the end user applications as well as cloud interaction among various physical and virtual entities in the ITS into one category. This layers also supports multi-cloud or edge cloud interactions supporting hybrid architectures, critical to distributed and mobile CPS domains such as smart cars where it is impossible to have all the entities associated with single central cloud. It is highly likely that each smart vehicle manufacturer will have its own cloud to support different applications to its customer, and in case data from one cloud is needed by applications in another cloud, negotiations and trust agreements must be established beforehand. As shown in Fig. 3.5 the interactions in this category include between end user applications and the cloud services (AP-CSR), within different cloud services (CSR-CSR) among different cloud (CL-CL) and edge/fog platforms (CL-FG) along with several indirect interactions as denoted by dashed boxes including applications and physical objects (AP-OB), applications and virtual objects of applications in the virtual layer (AP-VAP) among others. In addition, the three categories of the framework together support the indirect communication of user applications (AP) via the cloud services (CSR) interacting with the virtual object (VOB) to issue a command to physical sensor (OB) as represented by AP-CSR-VOB-OB. Similar counter part can be envisioned for clustered objects communication with the user applications.

So far, we discussed on the interaction among entities in the connected vehicle with the objects and applications belonging to different entities. However, interaction among different sensors and applications with a vehicle also need secure protection. This type of in-vehicle communication fits well into the aforementioned categories depending on the entities involved in the interaction which can be the physical sensors, virtual objects or the application in the vehicle supported by local edge gateway created in the vehicle itself and mediating all the interactions. CAN bus and other intra-vehicle communication can be protected by assigning ACLs and capabilities to ECUs to prevent spoofing and other attacks. TCUs or gateways have

---

[1] https://www.dds-foundation.org/what-is-dds-3/.

[2] https://www.w3.org/Protocols/rfc2616/rfc2616.html.

[3] https://coap.technology/.

been used to separate critical ECUs from non-important sub-networks and also act as a common external interface to connected car. Access control models should be developed for various interactions in each category to control communication and data exchange. Note that the proposed authorization framework does not include physical tampering and OBD port connectivity and hence been excluded from discussion.

3.3.2 Identified Access Control Approaches

Several research [4, 5, 7–9, 13–18] have worked in offering novel access control requirements and proposed different access control models and mechanisms supporting different enforcement architectures for CPS and IoT systems. An access control model for virtual objects communication was proposed by Alsehri and Sandhu [2, 13] offering CapBAC, ACLs and ABAC solutions to control interaction among cloud VOs. Bhatt and Sandhu [4] proposed AWS IoT access control model which provides a policy-based approach to control physical and shadow (virtual) communication along with devices which can connect with the IoT service. The mode uses MQTT based publish subscribe data exchange protocol, and devices must have the publish and subscribe permissions to enable data communication and support commands from applications and devices to other connected devices. In this regard, Gupta and Sandhu proposed dynamic groups and attribute-based solution for cloud assisted smart cars, and V2V solutions for trusted messages exchange using cloudlets. These works support external communication using cloud for application which can bear some latency, and edge based solution for real time application needs such as accident on road, or pedestrian crossing.

To our understanding *multi-level* policy decision and enforcement is needed, one for controlling the external communication, and another one to support in-vehicle communication of sensors. Access control solution controlling the external interface will enable authorized communication and access to vehicle's resources including sensors, data, ECUs and on-board applications from other external entities such as traffic lights, road side infrastructure, vehicles, pedestrian smart phones or cloud assisted applications. In case of the internal security mechanisms, deployed access control will limit the ECU's and in vehicle applications communication and messages exchange among different sensors controlled by CAN bus, or communication protocol using Bluetooth, ZigBee, WiFi or even 5G. It is to be noted that only deploying security at the external interface may not be sufficient to prevent adversaries, since they can impersonate as a trusted device and may be able to bypass external access control mechanisms. In addition, if some of the sensors and ECUs with the external interface are compromised, next level access control can protect and safeguard critical systems in the smart vehicles. In the ITS ecosystem, vehicles exchange messages with other vehicles and smart infrastructure as they move along on the road. Therefore, in ITS two types of data exchange scenarios can take place: static and dynamic. In static interaction, long term association among different data exchanging entities take place, such as the interaction among vehicle owner and the smart vehicle. This is an established relation, and chances of its change are very minimal. On the other side, dynamic communication is momentarily, for a shorter duration of time, and primarily happen because two smart entities share the location vicinity, or in some geographic range but have no prior history of data exchange and communication among them. In addition, static relationships are trusted and established which can share more personal information which is not true in dynamic interactions which are random and on the fly. Such relationships can also help us understand access control needs and foster new models for ITS.

Another approach which may require multi-level access control will depend on the type of operations requested on the objects, and the corresponding authority who may allow or deny that permission. For instance, someone requesting to control or issue engine repair operation on a smart autonomous vehicle will need permissions at both the owner and manufacturer level, whereas mechanic reading the tire pressure or engine fuel efficiency values may only need approval from the vehicle owner. We are convinced that clustered objects introduced in the E-ACO architecture are critical in determining access control decision. When two vehicles interact with each other, it is not actually the vehicles, but the sensors and application inside them sharing BSMs which they act upon autonomously. Hence, it can first be checked if the message has been shared from a trusted vehicle having valid certificates granted by central ITS authority. If such authorization is allowed at the initial level, next access control can be invoked for particular sensors, and applications which can access data from vehicles to make the final decision.

Trust can also be introduced among different interacting entities, which can be established based on the history of interactions, prior information shared (whether true or fake) or the relationship between them such as static or dynamic. For example, entities who have exchanged data can be considered to be more trusted as compared to random ones; vehicles with the same owner or among siblings are more trusted. The US Department of Transportation (USDOT) supported a Security Credential Management System (SCMS) which enables PKI based trust establishment to enable secure, confidential of BSM exchange and ensure integrity of V2V and V2I communication. In addition, attribute-based solutions can be deployed in ITS where different entities can inherit set of attributes from dynamic locations and geographic groups, or their current location, vehicle manufacturer or even the owner preferential policy. In such cases, policies can be specified using logic based languages which will determine and control sensors communication after establishing trust among the participating vehicles. These attributes can be used in access control decision determination which may include the current location, geographical position, speed and direction, acceleration, road surface temperature among several others. This can also support multi-level policies, at the cloud level to control V2X communications, and the local edge fog level to limit intra-vehicle communication among sensors and applications. Single central cloud, or multiple cloud edge architectures can support true ITS enabling both real time and diverse set of services offered by the ecosystem. With multi-cloud interactions also need access control and federated solutions. Apart from operational models, administrative models are also imperative to support the administration of ITS needs and deploy operational access control solutions.

3.3.3 Single and Multi-Cloud Cyber Physical Systems

In this section, we will explain some of the important use cases and scenarios which use the authorization framework discussed for ITS interaction scenarios encapsulating single cloud, multi cloud and hybrid cloud-edge/fog scenarios, as shown in Fig. 3.6. The classification of use case scenarios in single and multi cloud systems reflect the local and global distributed scope of different smart entities communication and various user applications. It must be noted that the applications envisioned in single cloud can easily be extended to multi-cloud systems and vice versa, with appropriate security deployments. The main motivation for us is to reflect different communication scenarios and data exchange which can happen in dynamic and distributed intelligent transportation systems, and the need for multi-layered access control decision and enforcement points.

Most of the application supported in ITS and smart cars are time sensitive and location centric requiring real time processing of the information which is exchanged among different smart vehicles, nearby restaurants, traffic infrastructures and different smart entities which are in the current location vicinity of the vehicle. In order to address the limitations of central cloud infrastructure with respect to latency and bandwidth, we also understand and firmly convinced that edge cloudlets will play a critical role, for storing as well as processing of messages to support ITS applications. In addition, vehicular cloud (VC) can be used which will harness the storage and computational resources of the vehicles and traffic infrastructure to create ad-hoc on-the-fly computing infrastructures. Therefore, single cloud applications can be supported by local edge or fog systems where virtual replicas of physical objects can be supported by edge or VC. In addition, each smart vehicle can also have in-built edge systems which can support in-vehicle communication. Different use case scenarios may require multiple virtual objects for each physical object which can be created either at the local VC or edge and also in central internet cloud. This will fulfill the requirements for a persistent state of objects or for applications which can bear some latency or in scenarios where the interacting connected vehicles and other traffic infrastructure entities are in the range of common cloudlet or a vehicular cloud. These use cases are elaborated in multi cloud and hybrid edge-cloud architecture scenarios as discussed below.

Single Cloud Applications and Access Control Single cloud applications consider smart objects which are interacting in a limited geographic area or associated with a single central cloud due to their characteristic (for example, belonging to the same manufacturer) and exchange relevant information. A pedestrian crossing road can send alert messages to nearby and approaching cars, or BMW 7 series remote parking capability can assist users with the ability to park it with a touch screen

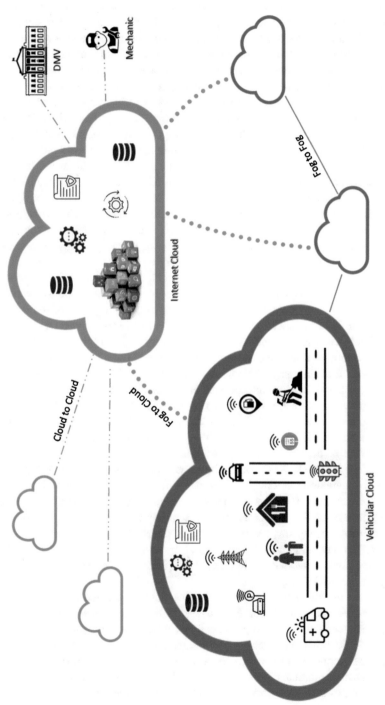

Fig. 3.6 Single and multi cloud hybrid smart cars and intelligent transportation systems

key using short-range communication. Nearby restaurant and gas stations can send coupons and discount offers to passing smart vehicles to increase their sales, or a traffic sign board detecting speed car and send a message to the vehicle which result in car seat vibration or even reducing the speed of a fully autonomous car are all supported with single cloud or edge cloudlets in a local limited geographic area. Entities at the physical layer such as vehicles, sensors or traffic lights will have virtual replicas or cyber entity (one to one or one to many) created in virtual object layer of E-ACO, which can be part of the central cloud or edge cloudlet. Topic or content based MQTT like publish subscribe protocols can be used to ensure the data from topics to which the entities are subscribed, which are published by entities from where different smart objects and applications want to get the data. The message broker deployed at the edge or cloud can ensure the reliability and complete mediation of the messages which are passed to only desired subscribers.

Apart from inter entity communication, in vehicle interactions are also supported in which ECU's and sensors along with in-vehicle applications exchange messaged with each other and also with the passengers sitting in the car. This intra-vehicle interaction is supported with the local edge or fog architecture created at each vehicle which also supports the creation of virtual objects of all the different sensors and ECUs in the vehicle. Critical ECUs such as used in the engine and transmission can be separated using gateway which can secure external interface of the vehicle. This also enables authentication and authorization to over-the-air (OTA) firmware updates from manufacturer and can enforce security policies for in-vehicle sensor and ECU communication. Access control enforcement points are required at three E-ACO layers including physical, virtual object and cloud services layer, where the communication and data exchange among authorized entities is allowed. V2V, V2I, V2P and all V2X interactions supported by DSRC, Bluetooth or WiFi etc. at the object layer requires authorized control to confirm confidentiality and integrity BSM and prevent malicious activities. Credit card information can be stored in a smart vehicle or can be attached to the cyber replica, which can support easy payment process on the toll roads. At the object layer, access controls are required for secure authorized interactions among ECUs sensors or user applications and clustered objects, for example, when a smart-phone is accessing infotainment applications or a plug-in user device into smart vehicle requires appropriate security controls. Access control mechanisms are required when physical objects interact with their virtual entities created in the edge or cloud. For example, the airbag ECU must be allowed to communicate with virtual entity to update state of the device or publish messages in the MQTT topic.

Our vision of ITS ecosystem supports virtual objects created for each physical object that will be critical for data and messages exchange among different heterogeneous objects. Virtual replicas will be created for smart objects and sensors inside the vehicle that can issue control commands to different connected vehicle. Hence, access control mechanisms are needed at E-ACO virtual object layer as well to control communication between virtual entities. Applications in the smart vehicles can also access on-board sensors, for instance, the lane-departure warning or tire-pressure monitoring system, which need to be authorized to ensure legitimate applications accessing data from sensors. Interaction between sensors and ECUs also require access control using the vehicle gateway. Attribute based access control solutions provide fine grained security policies and use environment and contextual factors to enable secure data exchange and interaction at both physical and virtual object layers. Therefore, in order to secure critical sensors and ECUs in vehicle first layer of access control will restrict external interfaces followed by in-vehicle access control offering second layer check supporting defense in depth approach.

Multi Cloud Hybrid Applications and Access Control   Several ITS applications and use case scenarios need multiple cloud and edge instances to provide services in broader geographic location, which may be real time or not sensitive to time. As an example, considering a vehicle manufacturer with a private cloud supporting all its vehicles, and can gather data from the entire fleet to perform diagnostic and analysis to find out potential run-time issues in the vehicles, and at the same time offer OTA firmware updates to increase the performance or fix bugs in the critical systems. Sometimes the information sent by the ECUs in the vehicle may need more than firmware update and needs immediate attention from a nearby mechanic to perform more in-depth diagnostics. It is very likely that the mechanic may have its own private cloud and may not have access to the vehicles' diagnostic data which is stored in the manufacturer cloud. In this case, trust must be established between these cloud platform to enable federation or collaboration of cloud data objects, which can then be shared with the mechanic cloud as on demand once approved by the car owner. In similar scenario, if a remote mechanic wants to push a message to sensor and ECUs in the car, cross cloud interaction is needed between the cloudlet or vehicular cloud which stored the virtual object of the sensor as well as the application the mechanic uses in his private cloud to push a message to the virtual shadow of the sensor and to the physical object. CarSpeak is an application which captures data from various sensors and object in the same car and also from various other associated vehicles, belonging to same or different cloud edge infrastructures. In that cases, such applications can have access to the virtual object across various edge cloudlets or vehicular cloud that will need trust among various edge and cloud service providers.

There could be possible scenarios where two or more vehicular clouds, cloudlets or hybrid vehicular cloud, fog and central cloud systems exchange data and information. As an example, consider a smart vehicle is about to reach the home destination of the driver, and the driver wants to turn on the thermostat to have the air conditioning ON. In this case, the user home may have its own local cloud and hub to support local communication which will also store all the cyber replicas of various smart objects inside it. In this case, cross cloud interaction of objects will take place in which application in the smart car will have to interact with the virtual replica of the thermostat which is associated with another cloud. In this scenario, as the home and car belong to the same owner, static trust can be established due to prolonged history of such interactions happening in the past as well. With trusted interaction, we may waive off the need for multi-level access control, and simply check trust as the only factor. In another case, consider the Department of Motor Vehicle (DMV) or local law enforcement pushes a broadcast notice regarding a stolen vehicle or some malicious activities detected in the city. In this case warning can be issued on the vehicle dashboard that will be displayed and alert the drivers. These services will be running in DMV private cloud or cloud supporting the police department, which will push messages to different vehicles in the city limits, which will also need multi cloud or hybrid edge cloud data access applications. DMV may also have dedicated cloud or edge infrastructure installed around the city using traffic infrastructure or highway which can exchange messages via cloudlets or cloud and can then forward it to nearby smart vehicles and appropriate smart sensors using cyber objects or WiFi communication within a geographic location.

Therefore, access control models and mechanisms are needed across single and multiple cloud hybrid architectures to enable secure communication among physical, virtual objects and services in ITS ecosystem.

3.4 Objectives of Proposed Frameworks

In this section, we discuss the objectives of our AC frameworks and relevant open research problems. The main objectives of developed authorization frameworks is to investigate and comprehend the access control focused security requirements and offer a vision to some of the open problems and access control oriented research directions. Here we highlight some research problems some of them will be addressed in the next few chapters of this book:

- **User-Based Device Authentication:** Some IoT devices such as one used in wearable or fitness smart watches are closely associated with individual who is wearing and about whom the information is being collected. Therefore, physical security of such devices is imperative, at the same time proper authentication mechanisms must also be deployed to ensure the person using is the legit user. This prevents stealing or loss of these devices where an adversary can compromise the integrity of the devices and the data in them. Several authentication mechanisms including fingerprint and heart rate have been discussed [19] and at the same time more research is needed.
- **User-Centric Data Security and Privacy:** Smart objects which are closely associated with users and collect data about their health, location and other PII information can potentially compromise their privacy. Therefore, users must have complete control and systems must be able to support user privacy preferences to what data a service can collect, and how the data would be secured in the cloud or third party services as well. A study [20] conducted on fitness tracker suggest several threats to user personal data due to inherent vulnerabilities in the smart devices at the same time provide some guidelines for users to define their needs. Similarly, connected environments like smart cars and ITS which support location based services to offer notifications and alerts to the users, must

support a user to decide which advertisements they want to have and filter out unacceptable ones.

- **Secure External Communication:** The smart connected objects are exposed to various external entities with no or limited prior history of interactions and data exchange, which broadens the attack surface. Smart vehicles and devices will have personal data which also require approval from the users and to confirm the privacy policies. In addition, the requirement to limit the data and information in critical sensors and ECUs, including who can issue control and commands to initiate different operations also require secure authorization. Smart entities and objects can be treated based on the established trust levels that will also determine how much data can be shared in comparison to random smart vehicles which exchange data for momentarily. A critical question to find answer is how such trust can be established between various objects in the ITS. In domains like ITS where the interaction is very dynamic and short lived, it will be challenging to deploy security policies. Hence it requires thoughtful insights and work from the community to deploy enforcement and policy models and architectures.

- **In-Vehicle Sensor and ECU Communication:** Messages in the vehicle supported by CAN bus or in-vehicle local gateway enable communication between sensors and ECUs. This gateway works as a firewall and limit all the critical sensors and ECUs in the vehicles from being compromised and exploited by remote adversaries. Authentication is needed to deter and prevent spoofing and false data injection in the sensors and ECUs. Data values captured by the ECUs must also be protected and over-the-air firmware updates must be secured. Several reports have illustrated the use of short range communication to vehicle's Bluetooth unit that can allow adversaries to gain access to vehicle objects, demands access control security. Physical tampering and direct OBD port access to ECU must be restricted.

- **Cross Cloud, Multi Cloud and Edge Cloud Interaction and Data Sharing:** Cloud and edge assisted IoT and CPS systems will need to support multiple cloud, edge or hybrid infrastructures. In order to ensure secure cross and multi cloud and edge communication, it is important to establish trust within different service providers that will also determine the level of information sharing and data exchange among trusted entities. IoT and CPS specific cross cloud and multi cloud access controls models and important security mechanisms need more focused attention. Similarly, gateway edge devices are needed for real time computation and for devices with capability constraints. Cloudlets [3, 8, 23] have been proposed for low bandwidth and latency requirements, which also need appropriate access control trusted cloud edge interactions.

3.5 Summary

In this chapter, we discussed and developed conceptual authorization frameworks for cloud assisted wearable IoT and connected cars ITS. These frameworks help us understand the different interactions happen among various entities in both of these critical domains, to help us determine the access control requirements, and contemplate existing or novel access control models needed to secure communication. These frameworks will act as guidelines for researchers to understand several access control decision and enforcement points necessary in the dynamic CPS ecosystem. Different communication and data exchange scenarios have been discussed followed by access control approaches in E-ACO layers. Real-world use cases with single and multi-cloud scenarios and access control requirements reflect the need and use of authorization framework for ITS and connected cars ecosystem. In the following chapters we will discuss the various access control models proposed and implemented in different CPS and IoT domains.

References

1. Alshehri, A., & Sandhu, R. (2016). Access control models for cloud-enabled internet of things: A proposed architecture and research agenda. In *IEEE International Conference on Collaboration and Internet Computing (CIC)* (pp. 530–538).
2. Alshehri, A., & Sandhu, R. (2017). Access control models for virtual object communication in cloud-enabled iot. In *2017 IEEE International Conference on Information Reuse and Integration (IRI)* (pp. 16–25). IEEE.
3. Bhatt, S., Patwa, F., & Sandhu, R. (2017). An access control framework for cloud-enabled wearable internet of things. In *2017 IEEE 3rd International Conference on Collaboration and Internet Computing (CIC)* (pp. 328–338). IEEE.
4. Bhatt, S., Patwa, F., & Sandhu, R. (2017). Access control model for aws internet of things. In *International Conference on Network and System Security* (pp. 721–736). Springer.
5. Gupta, M., & Sandhu, R. (2018). Authorization framework for secure cloud assisted connected cars and vehicular internet of things. In *Proceedings of the 23nd ACM on Symposium on Access Control Models and Technologies* (pp. 193–204).
6. Gupta, M. (2018). *Secure Cloud Assisted Smart Cars and Big Data: Access Control Models and Implementation*. Ph.D. thesis, University of Texas at San Antonio.
7. Gupta, M., Benson, J., Patwa, F., & Sandhu, R. (2019). Dynamic groups and attribute-based access control for next-generation smart cars. In *Proceedings of the Ninth ACM Conference on Data and Application Security and Privacy* (pp. 61–72).
8. Gupta, M., Benson, J., Patwa, F., & Sandhu, R. (2020). Secure V2V and V2I communication in intelligent transportation using cloudlets. *IEEE Transactions on Services Computing.* https://doi.org/10.1109/TSC.2020.3025993.
9. Gupta, M., Awaysheh, F. M., Benson, J., Azab, M. A., Patwa, F., & Sandhu, R. (2020). An attribute-based access control for cloud-enabled industrial smart vehicles. *IEEE Transactions on Industrial Informatics.* https://doi.org/10.1109/TII.2020.3022759.
10. Gupta, M., & Sandhu, R. (2021). Towards activity-centric access control for smart collaborative ecosystems. Preprint. arXiv:2102.11484.
11. Ouaddah, A., Mousannif, H., Abou Elkalam, A., & Ouahman, A. A. (2017). Access control in the internet of things: Big challenges and new opportunities. *Computer Networks, 112,* 237–262.
12. Sethi, P., & Sarangi, S. R. (2017). Internet of Things: Architectures, protocols, and applications. *Journal of Electrical and Computer Engineering, 2017*(Article ID 9324035), 25 p.
13. Alshehri, A., Benson, J., Patwa, F., & Sandhu, R. (2018). Access control model for virtual objects (shadows) communication for aws internet of things. In *Proceedings of the Eighth ACM Conference on Data and Application Security and Privacy* (pp. 175–185).

14. Gupta, D., Bhatt, S., Gupta, M., Kayode, O., & Tosun, A. S. (2020). Access control model for google cloud IoT. In *2020 IEEE 6th Intl Conference on Big Data Security on Cloud (BigDataSecurity), IEEE Intl Conference on High Performance and Smart Computing,(HPSC) and IEEE Intl Conference on Intelligent Data and Security (IDS)* (pp.198–208). IEEE.
15. Kaiwen, S., & Lihua, Y. (2014). Attribute-role-based hybrid access control in the internet of things. In *Asia-Pacific Web Conference* (pp. 333–343). Springer.
16. Hernández-Ramos, J. L., Jara, A. J., Marin, L., & Skarmeta, A. F. (2013). Distributed capability-based access control for the internet of things. *Journal of Internet Services and Information Security (JISIS), 3*(3/4), 1–16.
17. Gusmeroli, S., Piccione, S., & Rotondi, D. (2013). A capability-based security approach to manage access control in the internet of things. *Mathematical and Computer Modelling, 58*(5–6), 1189–1205.
18. Liu, J., Xiao, Y., & Philip Chen, C. L. (2012). Authentication and access control in the internet of things. In *2012 32nd International Conference on Distributed Computing Systems Workshops* (pp. 588–592). IEEE.
19. Nymi Band. https://nymi.com/. Accessed: 2017-01-08.
20. Fitness Trackers Could Benefit from Better Security. https://www.ed.ac.uk/news/2017/fitness-trackers-could-benefit-from-better-securit. Accessed: 2017-09-15.
21. Kumar, S., et al. (2012). CarSpeak: A content-centric network for autonomous driving. *SIGCOMM Comput. Commun. Rev,* 42(4) (Aug. 2012), 259–270.
22. Lee, U., et al. (2006). Mobeyes: Smart mobs for urban monitoring with a vehicular sensor network. *IEEE Wireless Communications,* 2006, 52–57.
23. Satyanarayanan, M., Bahl, P., Caceres, R., & Davies, N. (2009). The case for VM-based cloudlets in mobile computing. *IEEE Pervasive Computing, 8*(4).
24. Gupta, M., Patwa, F., & Sandhu, R. (2017). Object-tagged RBAC model for the Hadoop ecosystem. In *IFIP Annual Conference on Data and Applications Security and Privacy.* Springer (pp. 63–81).
25. Gupta, M., Patwa, F., Benson, J., & Sandhu, R. (2017). Multi-layer authorization framework for a representative Hadoop ecosystem deployment. In *Proceedings of the 22nd ACM on Symposium on Access Control Models and Technologies* (pp. 183–190).

Chapter 4
Access Control Models in Cloud IoT Services

4.1 Introduction

The proliferation of IoT and CPS is evident from the estimates that the global IoT market is expected to grow from \$916.9 billion in 2020 to \$1,994.3 billion in 2028. To cater this growing demand, various dominant and widely used cloud service providers, such as Microsoft Azure, Amazon Web Services (AWS), and Google Cloud Platform (GCP), have offered dedicated services to support different CPS and IoT use cases ranging from connected cars, farming to smart manufacturing and industrial IoT. The academic researcher community have developed several access control models, and at the same time, industry access control deployments have made significant strides. However, a consensus is still missing to understand and define access control solutions which will be able to fulfill the needs of industry as well as capture the abstractions and fine grained notions of access control developed in academia. To bridge this gap, it is important for academic researchers to understand the current deployments of access control in well-known and widely used cloud IoT platforms that will foster extensions to these deployments to propose novel integrative solutions. Microsoft Azure and Google Cloud platform have extended and adapted role-based access control (RBAC) for their cloud needs. Further, GCP IoT service uses RBAC solutions whereas AWS supports a policy-based solution to specify IoT access control requirements which are primarily to control publish and subscribe operation in MQTT protocol. Although, we can get some overview and have working knowledge of these access control solutions by reviewing vast documentation provided by these platforms, formal access control models are still missing which will not only help us fully understand the solutions but also help define new fine grained models to extend current capabilities.

In this chapter, we investigate and study AWS and GCP IoT services to understand authorization solutions and develop formal access control models for both of the dominant cloud services. These models have been abstracted based on the verbose documentation available in these platforms together with our hands-on

M. Gupta et al., *Access Control Models and Architectures For IoT and Cyber Physical Systems*, https://doi.org/10.1007/978-3-030-81089-4_4

demonstration of the access control capabilities to ensure the veracity and validity of our formal abstracted models. The AWS IoT access control model (referred to as AWS-IoTAC) is developed by extending the AWS Access Control (AWSAC) model proposed by Zhang et al. [1]. Similarly, for GCP IoT access control (referred as GCP-IoTAC), we will first develop the formal GCP cloud model which partially supports attribute-based access control (ABAC), followed by our formal definition which supports RBAC like access control solutions. Since the IoT service has new sets of subjects and objects with different types of operations, it is important to still ground our model on cloud capabilities and then extend them to IoT-specific needs. As we will discuss in the AWS IoT, this service partially supports ABAC with attributed for IoT devices, hence we will propose some extensions to fully integrate ABAC in AWS IoT for fine grained policies. We will further present different use cases including smart home IoT and E-health to demonstrate the security policies captured by our developed abstract models. In addition, we will also highlight the limitations of the access control capabilities of these platforms and propose some options of ABAC extensions to their current access control models to support flexible and contextual environmental attributes which will enable to support rich fine grained policies.

4.1.1 Chapter Organization

We will first discuss the formal AWS cloud access control (AWSAC) model and elaborate its definitions in Sect. 4.2. In Sect. 4.3, we will present and define the AWS-IoTAC conceptual model followed by formal definitions. We will also discuss the AWS-IoTAC mapping with the ACO architecture discussed in Chap. 2 and an elaborated smart-home use scenario utilizing the AWS-IoTAC model is demonstrated. Next, in Sect. 4.4, we will discuss the GCPAC and GCP-IoTAC models and present their formal definitions. E-health and smart-home use cases, which depict the GCP-IoTAC model, are discussed in Sect. 4.4.3. In Sect. 4.5, we will discuss limitations of the current state of the art in AWS and GCP IoT platforms, and propose fine grained extensions using attribute-based access control solutions. Section 4.6 concludes this chapter.

4.2 AWS Access Control Model

Amazon Web Services (AWS) is one of the largest cloud service providers that offers various services ranging from compute, storage, network, Machine Learning and Analytics, and Internet of Things (IoT). Initially, AWS utilized a policy-based access control approach, however, recently they have further expanded into attribute-based access control capabilities to enable fine grained access control in the cloud. Zhang

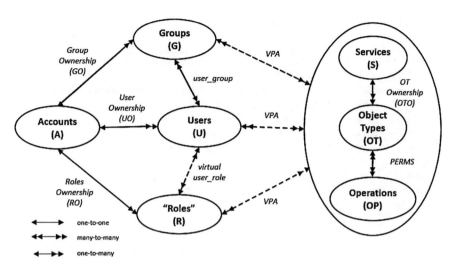

Fig. 4.1 AWS access control within a single account [1]

et al. [1] developed an access control model for AWS cloud services, known as AWSAC model.

Figure 4.1 shows the AWSAC model and its entities and relationships within a single account. We briefly describe the AWS Access Control (AWSAC) model and its formal definitions here, which in turn form a base for the AWS-IoTAC model presented in the next section. The AWSAC model within a single AWS account is shown in Fig. 4.1, with formal definitions presented in Table 4.1. AWSAC has seven components: *Accounts (A), Users (U), Groups (G), Roles (R), Services (S), Object Types (OT),* and *Operations (OP).*

In AWS, **Accounts** allows customers to contain their own cloud resources and also provides a standard billing unit for a customer based on their resource usage. The account generally belongs to the administrator of an organization who manages cloud resources for all the users within that organization. **Users** represent other individuals who can get access to cloud resources and these users within an account in cloud are generally created by the account holder and given access to required resources. To manage access and permissions for the users, **Groups** are defined which is a set of users. The **user_group** relation specifies the user to group assignment. The account holder can assign permissions to the group instead of every single user. For example, a DevOps team's access and permissions on resources can be specified by creating a group and assigning each user in the team to this group. Besides groups, there are also custom roles in AWS, known as **"Roles"**. These roles are different than the traditional roles as in the role-based access control. They are used for enabling inter-account access and establishing trust relationships between users and resources in different AWS accounts. *AssumeRole* action allows assigning roles to the users which enables these users to be trusted and allowed access to specific cloud resources through the **virtual user_role** relation. Since AWS roles

Table 4.1 AWSAC model components [1]

Definition 1
– A, U, G, R, S, OT and OP are finite sets of accounts, users, groups, roles, services, object types, and operations respectively
– User Ownership (UO): U → A, is a function mapping a user to its owning account, equivalently a many-to-one relation UO ⊆ U × A
– Group Ownership (GO): G → A, is a function mapping a group to its owning account, equivalently a many-to-one relation GO ⊆ G × A
– Role Ownership (RO): R → A, is a function mapping a role to its owning account, equivalently a many-to-one relation RO ⊆ R × A
– Object Type Ownership (OTO): OT → S, is a function mapping an object type to its owning service, equivalently a many-to-one relation OTO ⊆ OT × S
– PERMS = OT × OP, is the set of permissions
– Virtual Permission Assignment (VPA): VPA ⊆ (U ∪ G ∪ R) × PERMS, is a many-to-many virtual relation resulting from policies attached to users, groups, roles and resources
– user_group ⊆ U × G is a many-to-many mapping between users and groups where users and groups are owned by the same account
– Virtual user_role (VUR): VUR ⊆ U × R is a virtual relation resulting from policies attached to various entities (users, roles, groups), where users use *AssumeRole* action to acquire/activate a role authorized in VUR

are different than RBAC roles, they are represented as "roles" in Fig. 4.1. In the rest of the chapter, we use roles to signify "roles".

Within cloud architecture, there are several **Services** which consists of specific type of objects in it. The AWS cloud services hold the **Object Types** which represents a specific type of object. For example, in the compute service, known as Elastic Computing (EC), in AWS has a particular object, i.e. virtual machines (VMs). **Operations** represent authorized operations on the object types based on an access control policy attached to the object types or their owning services. AWS utilizes a policy-based access control (PBAC) mechanism. An AWS **policy** is a JSON file which specifies permissions on services and resources in the cloud. It comprises three main parts, also known as tags—*Effect, Action,* and *Resources*. There are also optional *Conditions* which allows to specify more restricted permissions. An access control policy in AWS can be attached to a user, a group, a role, or a cloud resource. **Virtual Permission Assignment (VPN)** is the process of virtually assigning permissions to users, roles, and groups through attaching policies to these entities. The policy can also be attached to a resource with a specific *Principal* which is an account, a user, or a role in that account. A single policy can incorporate multiple permissions, and a single entity can have multiple policies attached to it.

4.3   Access Control in AWS Internet of Things: AWS-IoTAC

In this section, we present a brief overview of AWS IoT service, a real-world cloud-enabled IoT (CE-IoT) platform, and also present a formal access control model for AWS IoT, known as AWS-IoTAC. This model is developed by extending the AWS cloud access control (AWSAC) model as discussed in the previous section. It is abstracted from extensive and dispersed AWS IoT documentation and hands-on investigation of this service to corroborate our understanding of the IoT service and its components.

AWS IoT is an IoT platform managed by one of the leading cloud service providers, Amazon Web Services (AWS). It enables secure communication between connected IoT devices and applications in the AWS cloud [2]. AWS IoT is a new service and has new IoT specific components besides the cloud access control components. An access control model for AWS IoT involves cloud entities, relationships, and new entities in the IoT space. It defines how these entities are authorized to communicate with each other securely.

4.3.1   Motivation

In a rapidly evolving and emerging IoT ecosystem, security and privacy are essential requirements to secure IoT devices, applications, and data as well as preserve users' data privacy. Access control models allow to prevent any unauthorized access on protected resources in a system. Within the IoT ecosystem, access and communication control are essential components for IoT security. Several access control models for IoT have been proposed recently [3–12, 32–36]. In addition, Ouaddah et al. [13] provide a comprehensive survey of IoT access control models. Most cloud providers including Amazon Web Services (AWS) [14], Microsoft Azure [15], and Google Cloud Platform (GCP) [16], have built and deployed their IoT services on top of their existing cloud services. These platforms generally utilize their cloud access control components for securing their IoT services. For instance, Azure and Google utilize a customized form of role-based access control (RBAC) [17, 18] with predefined roles and groups. GCP uses RBAC for its IoT solutions authorization [19] as well. On the other hand, AWS uses a policy-based access control (PBAC) approaches for its cloud and IoT services [2, 14, 37]. More recently, PBAC is also being adopted by other cloud IoT platforms, such as Azure cloud-IoT [20]. However, a formal access control model for real-world cloud-enabled IoT platforms is still lacking. It is essential to develop an abstract formal access control model [38] for a real-world CE-IoT that can capture all the IoT entities, relationships among them, and access/interaction associated with these entities and data access control collected by smart devices.

4.3.2   Formal Model and Definitions

Figure 4.2 shows the AWS-IoTAC model and its various components. AWS IoT
service is within the cloud, thus, the AWS-IoTAC model incorporates all of the
AWSAC components and relations along with new IoT components and relations.
The new or modified components and relations for AWS IoT are formally defined
in Table 4.2. A detailed description of these is presented here. The AWS IoT service
is shown as the **AWS IoT Service (AIS)** in the model which owns different entities
representing IoT devices, certificates and policies to enable secure authorizations
on IoT devices and associated entities. AIS is represented as a separate entity in the
AWS-IoTAC model to emphasize its importance and clearly depict other entities and
relations associated with it. The rectangular box is used to show the single existence
of the AIS service in AWS cloud.

With the AIS, there are **Certs (C)** (i.e., X.509 certificates [21]) which are issued
by a trusted entity, the certificate authority (CA). AIS generates the certs for IoT
clients (e.g., devices, applications, etc.) or accept certs created by these clients as
long as they are signed by a registered CA in AIS. The MQTT [22] based clients
widely use these certs to authenticate to AIS. MQTT is a machine-to-machine
(M2M) communication protocol that utilizes a publish/subscribe paradigm. **Devices
(D)** are a set of physical IoT devices, such as sensors, light bulbs, that can connect
to AWS IoT service. These devices exist in the physical space and are independent
of AIS, hence shown as blue in the model. In order for a device to authenticate to
AIS, a valid certificate (e.g., X.509) and its private key must be securely transferred
to the device. It also needs a root CA certificate issued from AWS to authenticate
and once it is authenticated to the AWS IoT service, a secure communication
channel is established between devices and AWS IoT service. There could be
different types of associations between IoT devices and certificates attached to
these devices. However, the AWS-IoTAC model considers one-to-one *cert_binding*
association between devices and certificates for better authorization management
and security. The *cert_binding* association is mutable such that an administrator
can update/modify the certificate association with a device based on its expiry or
revocation requirement. Certificates also play an important role in enforcing the
access control policies as access control policies are attached to certificates, which
enables enforcement of permissions on the physical IoT devices associated with
specific certificates.

IoT Objects (IO) represent virtual components of physical IoT devices in the
cloud. *Virtual objects* are virtual counterparts of real physical devices or standalone
logical entities, such as applications, in cyberspace [23]. In AWS IoT, a *Thing* and
a *Thing Shadow*, also referred as *Device Shadow*, are IoT objects. For each IoT
device, the model considers at least one *thing* along with its *thing or device shadow*
is created in cloud. The shadow for a device consists of a set of reserved MQTT
topics for enabling communications with the physical device through the shadow to
allow interaction with other IoT devices and applications, even when the device is

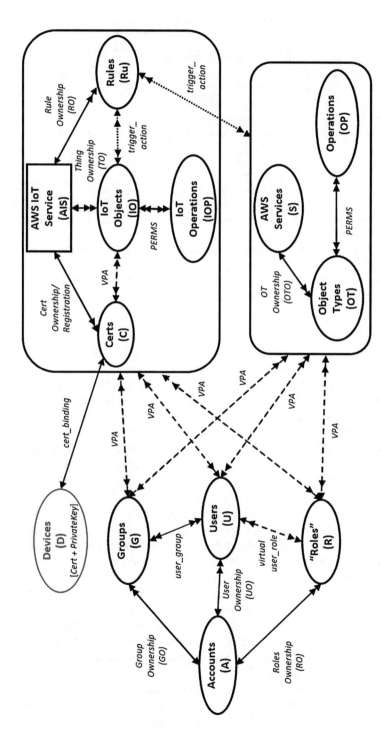

Fig. 4.2 AWS IoT access control (AWS-IoTAC) model within a single account

Table 4.2 AWS-IoTAC model—Additional components and relations

Definition 2
– AWS IoT Service (AIS) is one of the Services(S) in AWS
– C, D, IO, IOP, and Ru are finite sets of X.509 certificates, physical IoT devices, IoT objects, IoT operations, and rules defined in the rules engine of AIS respectively
– Cert Ownership/Registration (CO): C → AIS, is a function mapping a certificate to its owning service (AIS), equivalently a many-to-one relation CO ⊆ C × AIS
– Rules Ownership (RO): Ru → AIS, is a function mapping a rule to its owning service (AIS), equivalently a many-to-one relation RO ⊆ Ru × AIS
– Thing Ownership (TO): IO → AIS, is a function mapping the IoT objects to its owning service (AIS), equivalently a many-to-one relation TO ⊆ IO × AIS
– PERMS = OT × OP, is the set of permissions (including IoT permissions)
– Virtual Permission Assignment (VPA): VPA ⊆ (U ∪ G ∪ R ∪ C) × PERMS, is a many-to-many virtual relation resulting from policies attached to users, groups, roles, certificates, and resources
– cert_binding ⊆ C × D is a mutable one-to-one relation between X.509 certificate and IoT devices within a single account
– trigger_action ⊆ Ru × (IO × S) represents a many-to-many mapping between rules and IoT objects and AWS services on which a rule triggers action(s)

offline. The device shadow preserves the last known state of the IoT device and also there is a shadow document maintained which stores detailed data on device states.

There are mainly two types of IoT operations: *administrative*, such as create things, certificates, define policies, etc., and *operational*, which includes sending and receiving messages to perform specific tasks (e.g., turn on a device based on a sensor value). This model primarily focuses on operational aspects, therefore, **IoT Operations (IOP)** here are a set of operations defined for IoT service based on communication protocols used by devices and applications to communicate with each other and AWS IoT. There are four basic IoT operations for devices and clients using MQTT protocol: *iot:Publish* that allows devices/clients to publish a message to one or more MQTT topic(s), *iot:Subscribe* that allows a device to subscribe to one or more desired MQTT topic(s), *iot:Connect* enables an MQTT client to connect to the AWS IoT service, and *iot:Receive* enable a device/client to receive messages from subscribed topics. For devices or clients using HTTP(s), *iot:GetThingShadow* enables a device or client to get the current state of a thing shadow (its own shadow or other devices' shadows), *iot:UpdateThingShadow* allows a device or client to send messages to update/change the state of a thing shadow, and *iot:DeleteThingShadow* allows to delete a thing shadow. When a device or client sends a message to a virtual thing in the cloud, a new thing shadow is automatically created if it does not already exist. All of these operations will be allowed or denied based on access control policies defined for those devices or clients and resources in the model.

Besides IoT operations, there are also **Rules (Ru)** that can be defined in AWS IoT. These rules are SQL statements which trigger some actions based on the condition specified in the rule. For example, a rule receives data when there is a message sent to a device/thing given in the rule, then it checks if the defined condition is true (for example, if heartrate is greater than 120) and then one or more actions can be triggered. The actions could be to route the data from one IoT device to other IoT devices, or to other AWS services, such as AWS Simple Notification Service (SNS) to send sms messages to users if a certain condition is met in the defined rule. To authorize actions in rules, each rule is associated with a cloud IAM (Identity and Access Management) role that grants permissions to perform operations (e.g., publish) on IoT objects and/or on other AWS services. The **trigger_action** relation is a many-to-many mapping between rules and IoT objects and AWS services on which rules need to trigger one or more actions. AWS IoT service has its own specific access control policies with associated components (e.g., IoT operations, devices, and resources) and also utilizes cloud IAM policies to assign permissions related to IoT devices, IAM users, and IoT applications. IoT policy is similar to cloud IAM policy and is also a JSON with "Effect", "Action", and "Resource" keys and "conditions" if defined. The **Virtual Permission Assignment (VPA)** as discussed in the AWSAC model has been updated as per IoT service to incorporate IoT policies which are attached to X.509 certificates. The policy is enforced on the device through a device certificate attached to that device. A policy can be attached to many certificates, or many policies can be attached to a certificate.

In the AWS-IoTAC model, all the components and relations are defined within a single AWS account. However, with rapid enhancements of real-world cloud-enabled IoT platform, there are new capabilities being added to this platform. Here, we scope our discussion based on the AWS-IoTAC model published in [24].

4.3.3 AWS-IoTAC and ACO Architecture

Alshehri and Sandhu developed an Access Control Oriented (ACO) architecture for cloud-enabled IoT [25]. Here, we discuss a general mapping between AWS-IoTAC model and ACO IoT architecture. Figure 4.3 shows a mapping of different entities of the AWS-IoTAC model across different layers of the ACO architecture. For example, physical IoT devices map to the *Object layer*, and virtual IoT things or resources map to the *Virtual Object layer*. Similarly, AWS cloud services and resources are incorporated in the *Cloud Services layer*, and finally, users and applications are interacting with cloud and IoT devices at the *Application layer*. Users also communicate with physical devices at object layer, for instance, a user wearing a smart watch can interact with the watch and its application. Access control policies are defined for controlling access and operations on or from physical devices and applications (used by users). Overall, based on the mapping, we conclude that AWS-IoTAC aligns well with the ACO architecture.

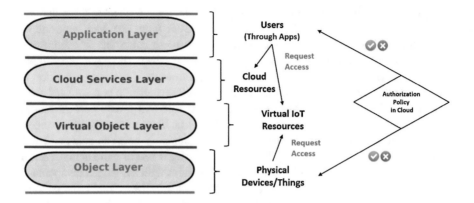

Fig. 4.3 AWS-IoTAC entities mapping to ACO architecture for cloud-enabled IoT

4.3.4 Use Case

This section presents a smart-home use case with two different scenarios including a smart thermostat and two light bulbs that are controlled through the AWS IoT service based on connected sensor inputs. The smart devices are connected to the AWS IoT platform which allows them to communicate with each other through the cloud. This use case mainly considers interactions between IoT devices, however, a more complex real-world example would involve multiple users and applications interacting with IoT devices. This use case demonstrates how the access control and authorization configurations are setup for enabling secure interactions between IoT devices based on the AWS-IoTAC model.

4.3.4.1 Use Case Setup and Configuration

Figure 4.4 shows an overview of the use case including connected devices, their respective virtual things/objects, and AWS Cloud and IoT Services involved in this use case. An AWS account was created to setup devices, things and thing shadows, rules, etc. in AWS IoT service and utilize other cloud services. For each physical device (two sensors, one thermostat and two light bulbs), a virtual object (*thing*) is created. A *thing* can have a *thing type* which allows to store specific configurations for things that are similar in nature, and *thing attributes* (key-value pairs) which represents properties of individual IoT devices. Suppose *Sensor_1* is of *Sensor* thing type and has two attributes *SType* (sensor type) and *Belongs* (belongs to). An administrator can set the values for these attributes while creating things in the AWS IoT service. In the AWS IoT console, X.509 certificates are created for each IoT thing using "one-click certificate creation" and then these certificates and private keys (public/private key pair) are copied on the physical devices manually by

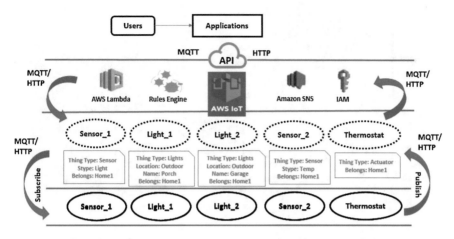

Fig. 4.4 Smart-home use case utilizing AWS IoT and cloud services

the administrator. Finally, we defined the access control policies for each IoT thing and attach them to their respective certificates which also enforces these policies on corresponding physical devices. The physical devices also need to have an AWS root CA certificate to be able to authenticate to AWS cloud. The CA certificate is used to verify identity of the server, viz., AWS IoT service here. An IoT device certificate is used to authenticate to the cloud and attached authorization policies during device authentication and once authenticated, access control aspects are enforced based on policies attached to the device certificate. In order to simulate physical devices (e.g., lights and thermostat device), we utilized AWS SDKs (Node.js) [26] provided by AWS, and simulated sensors as MQTT clients using MQTT.fx tool [22]. The communication protocol used for IoT devices and cloud interactions is MQTT, a publish-subscribe protocol over transport layer security (TLS).

There are two use case scenarios within the smart home use case presented in the following subsections. These scenarios utilize the *rules engine* to define *rules* in AWS IoT platform and trigger desired actions. The two actions triggered in these scenarios are invocation of a *Lambda function* and a notification sent to users by sending text messages through AWS SNS service. A cloud IAM "role" is associated with a rule to authorize it for accessing specific resources in AWS cloud and IoT services.

4.3.4.2 Use Case Scenarios

A. **Scenario 1:** Figure 4.5 shows scenario 1 which includes a temperature sensor and a smart thermostat, and a simple IoT policy as (a) and (b) in the figure respectively. A physical temperature sensor *Sensor_2* (which is shown as solid oval) senses home temperature and publishes a message with temperature data to its thing shadow which is shown as *Sensor_2* in Fig. 4.5a. A rule is also

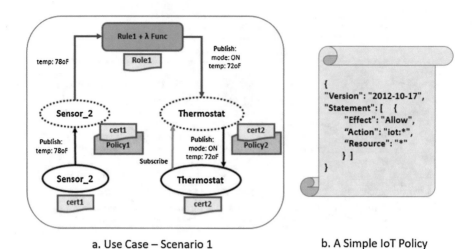

a. Use Case – Scenario 1   b. A Simple IoT Policy

Fig. 4.5 Smart-home use case scenario 1

defined which triggers a lambda function based on *Sensor_2* data. The lambda function publishes an update message to thermostat shadow for changing the state of *Thermostat*. For instance, if home temperature is greater than 78 degree Fahrenheit, then the rule invokes a lambda function which publishes a message on *Thermostat* thing shadow to turn it on with a desired temperature i.e., 72 degree Fahrenheit. The thermostat shadow syncs the desired state with physical thermostat (shown as solid oval) which has subscribed to its shadow topics for receiving update messages and changes its state. A simple authorization policy defined for both *Sensor_2* and *Thermostat* is shown in Fig. 4.5b. This policy allows any entity or device to do any IoT operation (e.g., publish, subscribe) represented as "iot:*" in policy on any resource ("*") in AWS IoT. It is attached to *Sensor_2* and *Thermostat* certificates that are attached to things and copied onto corresponding physical devices. Therefore, as per this policy, physical IoT devices can perform any IoT operation on any resource in AWS IoT. It is a simple and flexible policy but for enabling more fine grained access control, we may need to define a more restricted policy as discussed in Scenario 2.

B. **Scenario 2:** Here, we discuss a comprehensive scenario with a fine grained authorization policy as shown in Fig. 4.6. *Sensor_1* is a light sensor that monitors light level of its surrounding environment and if light level is low (i.e., it is dark), then a message is published to turn on outdoor lights which are *Light_1* and *Light_2*. As soon as the lights get turned on, home owner receives a text message notification about the state change of the lights. In this scenario, a more restrictive access control policy is defined for *Sensor_1*. The policy is based on *Sensor_1* thing attributes specified as a *Condition* in the policy. Figure 4.6b shows the fine grained policy which includes two policy statements—first statement authorizes a client to connect to AWS IoT only if its client ID is *Sensor_1*, and second

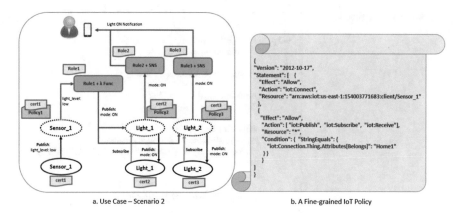

a. Use Case – Scenario 2   b. A Fine-grained IoT Policy

Fig. 4.6 Smart-home use case scenario 2

statement allows publish, subscribe, and receive operations on any resources if and only if the client requesting access has a thing attribute *Belongs* and its value is *Home1*. This policy employs thing attributes which represent the characteristics of IoT things/devices and allow to define more fine grained access control policy as shown in Fig. 4.6b.

AWS IoT policies allow utilizing thing attributes only for those clients (devices/things) that request access on resources, i.e., source clients or devices. However, it is also necessary to consider attributes of target resources on which IoT operations need to be performed. For example, we want *Sensor_1* to publish data only on those lights which have an attribute *Location = Outdoor* and may also need to use other attributes to consider the time of the day. While AWS-IoTAC model was developed, AWS IoT could not utilize attributes of target things/devices and/or clients in IoT policies. To demonstrate the importance of both source and target device/client attributes, we utilized rules and lambda functions as shown in Lambda function code snippet in Fig. 4.7. This piece of code shows how it searches for things (and their corresponding physical devices) that have an attribute (key and value), *Location = Outdoor*, and get a list of those things, i.e., *Light_1* and *Light_2* in this use case scenario. A JSON update message is published on these lights' shadows update topic, which then is sent to the physical devices to sync and update their states with shadow updates. Once the lights are turned on, a text message notification is sent to home owner as per the rules, *Rule_2* and *Rule_3*, through the AWS SNS service.

Fig. 4.7 Lambda function

```
...
var params2 = {attributeName: 'Location',
               attributeValue: 'Outdoor'
};
iot.listThings(params2, function(err, data) {
    ...
    for (i in data.things) {
        x = data.things[i].thingName;

        var params3 = {
        topic: '$aws/things/'+x+'/shadow/update',
        payload: new Buffer('{"state": {"desired" : {"light" : "ON"}}}'),
        qos: 0
        };

        iotdata.publish(params3, function(err, data){
            ...
```

4.4   Google Cloud Platform Access Control Model

Google Cloud Platform (GCP) is another leading cloud service providers which has also introduced their IoT service and utilizes the cloud access control and IAM capabilities. However, a formal abstraction of GCP cloud and its IoT service has been lacking. Therefore, GCP cloud and IoT access control models are developed in [8] based on the GCP documentation and hands-on experimentation with real-world use cases in GCP cloud and IoT platforms.

4.4.1   GCP Access Control (GCPAC) Model

This section presents a formal access control model for GCP cloud services, also known as GCPAC, along with its formal definitions. Figure 4.8 shows the GCPAC model within a single project context. Table 4.3 presents formal definitions of the GCPAC model. The main elements of GCPAC are Organizations (O_{rg}), Projects (P), Users (U), Groups (G), Roles (R), Services (S), Object Types (OT), Objects (OB), and Operations (OP).

Organizations and Projects: In GCP, cloud resources are administered by resource manager and are organized into hierarchical structures. The organization (O_{rg}) is the root node in Google resource hierarchy and it is also a super node for the project (P). It offers a general overview of different projects and project specific resources. It is automatically generated for a user (U) or a group (G) of users who are associated with a specific domain. The relationship between an organization and its projects is known as *Project Ownership (PO)* which is shown as a many-to-one relationship in Fig. 4.8. An organization can own multiple projects, and a project is owned by an organization. The model presents access control and authorization

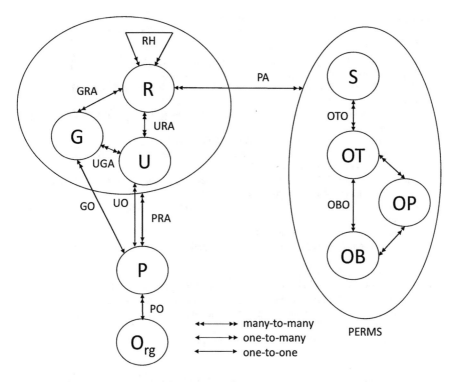

Fig. 4.8 GCP access control model within a single project

aspects between different cloud services and resources within a project. All the resources, such as services, their object types, and objects, are associated with a specific project in an organization in GCP.

Users and Groups: Users (U) can access various services in GCP. These users can be identified through their email address, such as user002@gmail.com, or by a service account (represented as service-124426731641@gcp-sa-cloud iot.iam.gserviceaccount.com in GCP). Besides a cloud user, there is also a service account entity in GCP, associated with a user, which permits inter-service access between two cloud services. Within a project, there can be multiple users and groups (G) of users. The groups are represented as gcp001@googlegroups.com which consists of multiple users (group of user email addresses represented as google groups). The assignment relation between users and groups is defined as *User-Group Assignment (UGA)*. Groups enable administrators to assign access control permissions to multiple users at once and also simplify updates or modifications of those permissions.

Roles: A role incorporates a set of permissions which can be assigned to groups and users in GCP. *User-Role Assignment (URA)* represents association of a user and a role, and *Group-Role Assignment (GRA)* represents association of a role and a

Table 4.3 Formal GCP access control model definitions

Basic sets and functions

- O_{rg}, P (finite set of organizations and projects respectively)
- U, G, R (finite set of users, groups, and roles respectively)
- S, OT, OB, and OP (finite set of services, object types, objects, and operations respectively)
- Project Ownership (PO): P $\rightarrow$ O_{rg}, is a function mapping a project to its owning organization,
  equivalently a many-to-one relation PO $\subseteq$ P $\times$ O_{rg}
- User Ownership (UO): U $\rightarrow$ P, is a function mapping a user to its owning project,
  equivalently a many-to-one relation UO $\subseteq$ U $\times$ P
- Group Ownership (GO): G $\rightarrow$ P, is a function mapping a group to its owning project,
  equivalently a many-to-one relation GO $\subseteq$ G $\times$ P
- Object Type Ownership (OTO) : OT $\rightarrow$ S, is a function mapping an object type to its owning service,
  equivalently a many-to-one relation OTO $\subseteq$ OT $\times$ S
- Object Ownership (OBO) : OB $\rightarrow$ OT, is a function mapping an object to its owning object type,
  equivalently a many-to-one relation OBO $\subseteq$ OB $\times$ OT
- directUG : U $\rightarrow$ 2^G, mapping each user to a set of groups, equivalently UGA $\subseteq$ U $\times$ G
- directUR : U $\rightarrow$ 2^R, mapping each user to a set of roles, equivalently URA $\subseteq$ U $\times$ R
- directGR : G $\rightarrow$ 2^R, mapping each group to a set of roles, equivalently GRA $\subseteq$ G $\times$ R
- PERMS= $2^{S \times (OT \cup OB) \times OP}$, set of services, object types, objects permissions
- Role Hierarchy (RH): RH $\subseteq$ R $\times$ R, partial order relation $\succeq_r$ on R

Assignments

- Project-Role Assignment (PRA): PRA $\subseteq$ P $\times$ (URA $\bigcup$ GRA),
  a many-to-many organization to project-role pair assignment
- Permission Assignment (PA): PA $\subseteq$ R $\times$ PERMS,
  a many-to-many permission to role assignment relation

Effective roles of users

- effectiveR: U $\rightarrow$ 2^R, defined as effectiveR(u) $= \cup_{g \in directUG(u)}$ directGR(g) $\bigcup \cup_{r \in roles(u)}$
  $\{p|[(\exists\, r'' \leq r)\, (p, r'') \in PA]\}$
  where role(u) $\subseteq \{r \mid (\exists\, r' \geq r)\, [(user, r') \in URA]\}$

---

group. The permissions within a role specify what actions are allowed on google cloud resources. GCP utilizes a customized RBAC mechanism, where permissions over cloud resources are expressed in a policy file, which is attached to specific roles. Once a role is assigned to an entity (a user or a group), then it gets all the permissions associated with that role. There are three types of roles in GCPAC model: **Primitive roles**, **Predefined roles**, and **Custom roles**. The **Primitive roles** comprises the *Owner, Editor,* and *Viewer* roles which can be assigned on a project. In addition to **Primitive roles**, there are **Predefined roles** that provide granular access on particular resources in a cloud service. Both primitive and predefined roles are fixed, whereas **Custom roles** offer granular access based on a user-specified list of permissions, and are adjustable to meet users' requirements. Roles can also be controlled by an optional *condition* feature that allows permissions on specific resources as long as certain configured conditions are met. The conditions are defined in the role binding with a resource's policy that comprises of users, roles and conditions. For example, one user can be assigned to one or more roles, and every role binds to zero or more conditions. A condition contains title, description (optional), and expression. Two kinds of condition attributes are provided: **Resource attributes** (resource types, resource name), **Requested attributes** (date, time). For instance, resource.name.startsWith("projects/develop001/buckets/bucket-001") represents a resource whose name must start with the provided value. In GCP, some of the services support condition feature, such as Cloud Storage—Buckets, Compute Engine, Identity-Aware Proxy, Cloud Key Management Service, and Resource Manager. However, in GCP, every policy has a maximum of 100 conditional role bindings, and only 12 logical operators are allowed in a condition expression.

Services: GCP [27] offers many cloud services including Compute Engine, Storage, Big Data, IoT, and Machine Learning, etc. The cloud IAM regulates access to these services and their object types and objects. The GCPAC model in Fig. 4.8 shows that permissions to access GCP services and their object types and objects are assigned through cloud IAM (using roles and their policies). While most of the GCP services utilize RBAC for access control needs, some services also utilize other mechanisms with roles. For example, access control in GCP storage service is based on roles and access control lists (ACLs) which define permissions on specific buckets in the storage service. **Object Types and Objects**: Within a cloud service, there exist different types of objects. Object Types (OT) indicates a certain type of an object in a cloud service. For example, in Compute Engine, instances (virtual machines (VMs)) and registry are the object types and a specific instance (VM) is an object of that service.

Operations: There are several operations (OP) that can be performed on a cloud service, its object types and objects. These operations are specifically based on the service and its object types and objects. In compute engine, operations on instances can be create, update, and delete that are provided to this service for performing desired actions on instances. Similarly, other services have specific operations provided for them and their object types and objects. Next, we have *directUG* function which represents a mapping of a user to one or more groups as presented in Table 4.3. In addition, both users and groups can be assigned to one or more roles and represented by functions *directUR* and *directGR* respectively. Based on the role assignments to entities in the model, actions on services, objects types and objects are authorized by these roles that each entity has with a set of associated role permissions.

PERMS: These are a set of permissions on specific cloud services, their object types and objects. It is represented as a power set of the cross product of GCP Services (S), Object Types (OT), Objects (OB), and Operations (OP). Permissions can be defined for any entity within a service and the service itself which allows operations in that service. Policies attached to roles permit or deny operations on the object types and objects themselves.

Role Hierarchy (RH): It is a partial order relation on role, written as $\succeq_r$. Within a service, when a role is assigned to an object type, then role permissions are inherited by all the objects of that object type. With role hierarchy, roles assigned to users and groups in a project are inherited and applied on all resources within that project. As shown in Table 4.3, the effective roles of users are represented as *effectiveR*, where users and groups (and users within a group) get effective permissions on various resources in a project indirectly through role hierarchy. The process of assigning permissions to users and groups through a role in a project is called *Project-Role Assignment (PRA)*. Also, the process of assigning permissions over roles to get authorization for accessing services, object types, and objects is called *Permission Assignment (PA)*.

4.4.2 Access Control in GCP Internet of Things

GCP-IoT is an Internet of Things (IoT) platform that contains Cloud IoT Core (CIC) service along with other GCP services. This platform enables secure communication among IoT devices, users, applications, and services in the GCP cloud. An access control model for GCP-IoT requires diverse entities in the IoT space, and it should also formally define how these entities communicate securely with each other. GCPAC model acts as the base model for developing GCP-IoT access control model, thus, entities in cloud model are integrated with new IoT components to develop the GCP-IoTAC model. GCP-IoT is based on scrupulous research of the comprehensive documentation on GCP-IoT as well as examination of this service to validate our understanding of GCP-IoT service [27]. Figure 4.9 shows the GCP-IoTAC model with various components, including all GCPAC components and relations as well as new components and relations related to the CIC service and other basic services. Table 4.4 shows formal definition of modified and new additional components and relations. There are extra nine components in the GCP-IoTAC model: Cloud IoT Core (CIC) service, cloud Pub/Sub (PS) service, Cloud Functions (CF) service, IoT Devices (D), Virtual Devices (VD), Registries (R_G), Topics (T), Functions (F), and IoT Operations (OP_{IoT}).

CIC Service: Cloud IoT Core (CIC) is a fully managed IoT service that permits us to securely connect, manage, and collect data from various IoT devices. In conjunction with other services on Google Cloud IoT platform, CIC service offers a comprehensive solution for gathering, processing, and analyzing IoT data in real-time to assist and improve operational efficiency. Two basic components of CIC service are: a device manager and MQTT/HTTP protocol bridge. They help in registering devices and connecting a device to GCP respectively. Other main services that CIC service collaborate with are *cloud Pub/Sub (PS)* service and *cloud functions (CF)*. PS is an asynchronous messaging service, that utilizes two levels of interaction between the publisher and subscriber, while CF is a serverless implementation environment for connecting cloud services. Publisher is a role, and

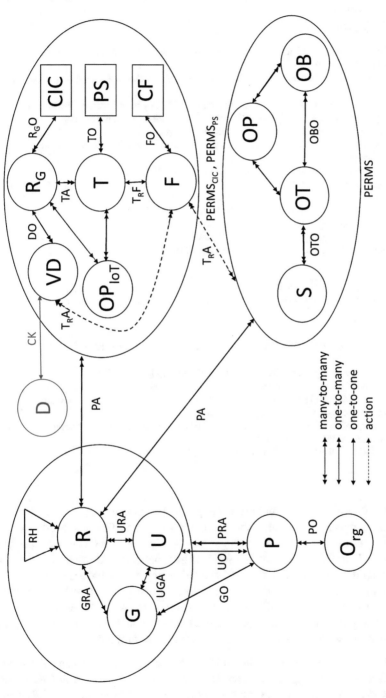

Fig. 4.9 GCP-IoT access control model within a single project

Table 4.4 Formal GCP-IoTAC model—Elaborating additional components with respect to GCPAC model

Basic sets and functions

- CIC, PS, CF (Cloud IoT Core, Cloud Pub/Sub, and Cloud Functions are one of the Services in GCP)
- D, VD, R_G (finite set of physical IoT devices, virtual IoT devices, and registry respectively)
- T, F, and OP_{IoT} (finite set of topics, functions and IoT operations respectively)
- Cryptographic Keypairing (CK): D → VD, is a one-to-one relation between physical IoT devices and virtual IoT devices within a project
- Registry Ownership (R_GO): R_G → CIC, is a function mapping a registry to its owning service (CIC),
  equivalently a many-to-one relation R_GO ⊆ R_G × CIC
- Topic Ownership (TO): T → PS, is a function mapping a topic to its owning service (PS), equivalently a many-to-one relation TO ⊆ T × PS
- Virtual Device Ownership (DO): VD → R_G, is a function mapping a virtual IoT device to its owning registry,
  equivalently a many-to-one relation DO ⊆ VD × R_G
- Function Ownership (FO): F → CF, is a function mapping a function to its owning service (CF),
  equivalently a many-to-one relation FO ⊆ F × CF
- $PERMS_{CIC} = 2^{CIC \times R_G \times OP_{IoT}}$, set of Cloud IoT Core and registry permissions
- $PERMS_{PS} = 2^{PS \times T \times OP_{IoT}}$, set of cloud Pub/Sub service and topic permissions

Assignments

- Topic Assignment (TA): TA ⊆ R_G × T, a many-to-one topic to registry assignment relation
- Permission Assignment (PA): PA ⊆ R × ($PERMS_{CIC}$ ⋃ $PERMS_{PS}$)
  a many-to-many permission to role assignment relation
- trigger_function (T_RF) ⊆ T × F represents a one-to-one mapping between topics, and functions on which a function triggers
- trigger_action (T_RA) ⊆ F × (VD × S) represents a many-to-many mapping between functions and virtual IoT devices,
  and GCP services on which a function triggers action(s)

entities with Publisher role publish messages to a particular topic, and subscriber is also a role, which allows entities to get messages on a particular subscription. The two services are presented in rectangle box in Fig. 4.9 to indicate their presence in GCP-IoT. In GCP-IoT, Cloudiot.serviceAgent handles CIC resources, involving publishing data to cloud Pub/Sub. Device telemetry data, which is event data like temperature measurement or other data from sensors, is redirected to a cloud Pub/Sub topic and triggers CF to send update configuration. Telemetry data or a device's current state are sent from a device to cloud, and device configuration, such as light_mode = ON, is sent from cloud to the light to turn it on.

Devices: A device is a processing unit that collects data and exchanges data with the cloud by connecting to the Internet. Devices are not involved in CIC service, so they are shown in a different color in Fig. 4.9. **Virtual Devices and Registry**:

A legitimate "provisioner", who is usually a user configuring a device, is expected to have generated a registry in GCP-IoT, and have permissions to generate virtual devices within the registry. In GCP-IoT, a virtual device is considered for every physical device. A virtual device within the registry is known as an *Object* in cloud, and registry is known as an *Object Type*. Virtual devices are digital representations of actual physical devices, or could be an entity or process like applications in the digital space. For each data generating virtual device, there is a *Cryptographic Keypairing (CK)*, and each device also generates a JSON Web Token (JWT) signed with its private key while connecting to CIC. The private key is authenticated by CIC using the device's public key. There can also be an expiration date set for the public key but is optional while creating a self-signed X.509 certificate. A registry is a group of virtual devices.

Topics and Functions: Topic represents an ingest of IoT data which flows through the PS service. For receiving messages that are published by other devices or applications, subscriptions are created on a particular topic to receive those messages. Telemetry event data is published on a topic through a service agent for Cloud IoT service. A service agent is an entity that has a role to allow interaction between Cloud IoT Core and Pub/Sub. These service agents provide a mechanism to allow inter-service interaction between different services, i.e., CIC, PS, and CF. These agents have appropriate roles assigned to them with required permissions. Another two service agents, such as *Cloud Dataflow Service Agent*, and *Cloud Functions Service Agent*, are generated automatically to allow communications across other services. When a message is published on a topic by Cloud Functions Service Agent, then a function is triggered to do specific actions.

IoT Operations: There are IoT specific operations defined for IoT service. These operations can be classified as per the communication protocol used by IoT devices, such as MQTT or HTTP. IoT operations on the registry are: *connect, receive, sendCommand, updateConfig*, and the operations that can be performed on the topic are: *publish, subscribe*. Similarly, for cloud functions, there is an *invoker* operation.

$PERMS_{CIC}$ is the set of cloud IoT service and registry permissions. It is represented as a power set of the cross product of *CIC*, registries (R_G), and operations (OP_{IoT}). $PERMS_{PS}$ is the set of cloud PS service and topics permissions. It is represented as a power set of the cross product of cloud PS, topics (T), and IoT operations (OP_{IoT}). The permissions on cloud functions are assigned by PS and CIC.

When a registry is generated in PS, a topic is created for publishing telemetry events and data. *Topic Assignment (TA)* allocates one or many topics to the registry. All devices are created in the registry and get permissions to publish the telemetry data through these required permissions– $PERMS_{CIC}$ and $PERMS_{PS}$. On the other hand, devices interaction control is managed by **allow** and **block** actions provided in the Google console. Events such as publishing a message to a Pub/Sub topic triggers background cloud functions, which are triggered by *trigger_functions* ($T_R F$) as shown in Fig. 4.9. These functions work well with small amount of data, however, Cloud Dataflow service is used when a large amount of data is

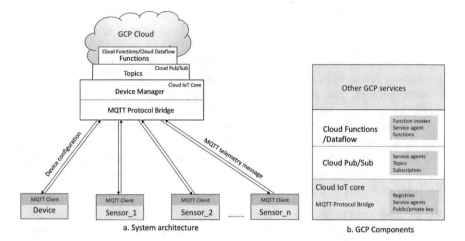

Fig. 4.10 GCP cloud use case implementation architecture

published. Finally, *trigger_actions* ($T_R A$) is a many-to-many relation between functions, virtual devices, and GCP services where the functions trigger action(s).

4.4.3   E-Health Use Case

Figure 4.10 shows the use case implementation architecture in GCP. Devices and various GCP cloud services and components are used in the proof-of-concept implementation of the use case. E-Health use case is presented in this section which includes smart medical devices, sensors, and GCP cloud and IoT services.

IoT devices and sensors collect data from user and surrounding environment and utilize this data to complete desired tasks. IoT devices were simulated as virtual machines (VMs) which included a MQTT service that allows to send and receive telemetry data through MQTT protocol bridge. Theses devices and sensors are authenticated by *Cloud IoT Core*. Once they have been authenticated, MQTT protocol bridge enables data from these sensors and devices to be forwarded to *Pub/Sub* service, as depicted in Fig. 4.10a. The cloud *Pub/Sub* is a robust message queuing system capable of handling streams of messages, and *Cloud Functions* is a cloud service that can trigger actions to send configuration changes on or send messages to specific devices. Moreover, there is another service, *Cloud Dataflow*, that transforms, enriches, and stores telemetry data using distributed data pipelines. This service is integrated into GCP platform, similar to Cloud Firestore and Cloud Bigtable services used for storage functions. Figure 4.10b depicts specific GCP components. In *Cloud IoT Core*, public-private keys need to be created for devices or sensors which enable them to securely authenticate and then send and receive data. Similarly, device registries are created in this service, and MQTT protocol bridge resides here. The MQTT protocol bridge acts as a server for IoT devices to enable bidirectional communication.

In this use case, devices and sensors collect patients' data, and healthcare practitioners monitor patients' data through IoT applications using GCP-IoT service. This use case demonstrates how smart devices interact with different types of users through cloud services. IoT devices, such as smart watch and heart rate sensor, collect user (patient) data and assist in reducing the number of a patient's visits to the hospital. In addition, such E-health devices and applications allow any user to monitor their health and provide predictive analytics based on collected data.

4.4.3.1   Use Case Setup and Configuration

E-health use case setup is shown in Fig. 4.11. There are various sensors–a body temperature sensor, a heart rate sensor, an EMG sensor, and a wearable device, and users–doctor, nurse, etc. in this use case. For each physical device, there is a virtual device in IoT registry. First, we create a project and an administrative role is given by default. Besides, a cloud IoT provisioner role can also be assigned to other users for creating a registry and virtual devices in cloud. A user with Provisioner role creates a registry and virtual devices in it. Furthermore, there are topics created in the registry. Besides, for different types of users in this use case, Cloud IAM manages users' roles. For monitoring a user's health, sensors associated with a user publishes the data to the telemetry feeds through cloud Pub/Sub topic. Once there is data on a topic, it triggers actions such as sending notification to doctor and device user based on predefined conditions. The policy for a topic–topic1 is shown in Fig. 4.12. The policy in Fig. 4.12 shows that the inter-service access are managed by service accounts and their roles have necessary permissions to access the resources in cloud. For storing the user's health data, buckets are used with a condition on firebase rules.

Next, we discuss how access control mechanisms in this use case are designed and configured based on the GCP-IoTAC model.

4.4.3.2   Description of E-Health Use Case

Figure 4.13 summarizes communications between different components of the E-Health use case, i.e., physical IoT devices, virtual devices, device users, and hospital practitioners. To simplify the sequence diagram, here it presents only one *body temperature sensor* and interactions with different users and services. This sequence diagram shows two parts: (i) authentication, and (ii) authorization. First the *body temperature sensor* authenticates in cloud-IoT service. For authorization, CIC service provides a role "IoT Cloud Provisioner" that has permission to create virtual devices in the registry. Thus, a user with provisioner role can create virtual *body temperature sensor* with unique-ID and generate a public-private key pair. The public-private keys allow a device to authenticate when it tries to connect to cloud. The device manager registers *body temperature sensor*, and notifies the provisioner.

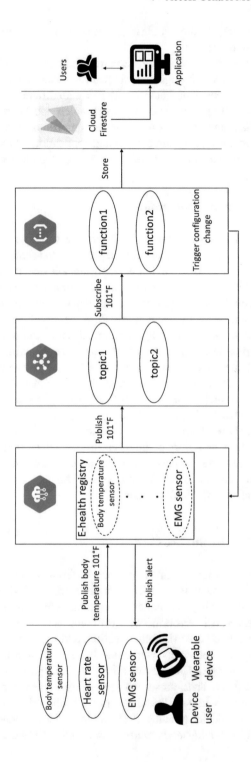

Fig. 4.11 E-health use case setup

```
"bindings": [
  {
    "members": [
      "serviceAccount:service-124426731641@gcf-admin-    ◄─────   Service
)bot.iam.gserviceaccount.com"                                     account
    ],
    "role": "roles/cloudfunctions.serviceAgent"          ◄───────   Role
  },
  {
    "members": [
      "serviceAccount:service-124426731641@gcp-sa-
oudiot.iam.gserviceaccount.com"
    ],
    "role": "roles/cloudiot.serviceAgent"
  },
  {
    "members": [
      "serviceAccount:service-124426731641@dataflow-service-producer-
rod.iam.gserviceaccount.com"
    ],
    "role": "roles/dataflow.serviceAgent"
  },
  {
    "members": [
      "serviceAccount:service-124426731641@firebase-rules.
ιm.gserviceaccount.com"
    ],
    "role": "roles/firebaserules.system"
  },
  {
    "condition": {
      "expression": "resource.name.startsWith(\"projects   ◄───────   Condition
leveloping-81990/zones/us/buckets/deep12\")",
      "title": "condition001"
    },
    "members": [                                                     Google
      "user:user001@gmail.com"                             ◄──────   account user
    ],
    "role": "roles/storage.admin"
  }
],
"etag": "BwWeNuqNVAA=",
"version": 3
```

Fig. 4.12 Policy of topic1

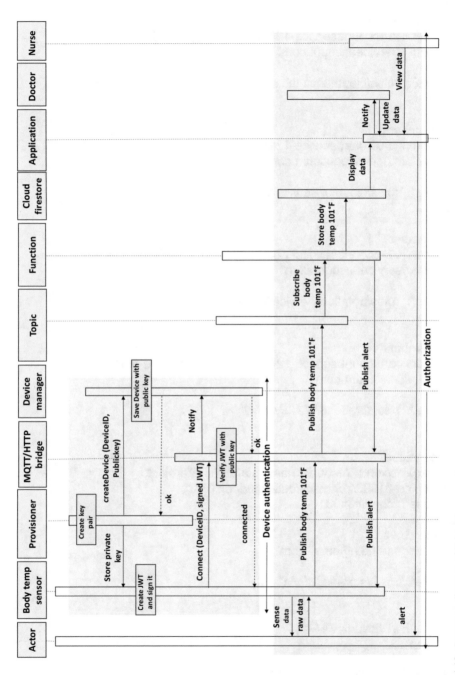

Fig. 4.13 Sequence diagram of E-health use case

The authentication flow works as follows. The *body temperature sensor* generates a JSON Web Token (JWT) signed with the private key. The MQTT bridge verifies this JWT against the device's public key and accepts the connection and notifies the device manager. The connection is open until the JWT expires.

The *body temperature sensor* senses the body temperature of a user and publishes its telemetry data to a topic in cloud through MQTT/HTTPS bridge. An important thing here is that roles are assigned on the registry, not on the device, for publishing messages, and any role cannot be assigned on a virtual device. Besides, communication of virtual *body temperature sensor* can be controlled through **allow/block** options available in cloud. Cloud IoT Service Agent manages publish operation on *topic1* using cloud Pub/Sub service. Once a message is published on associated topic, a background *function1* triggers indirectly when a message is published on *topic1*. If the body temperature is greater than 100 degree Fahrenheit, then *function1* fires a callback alert, which contains the registry and device IDs. Cloud Functions Service Agent subscribes to topic1 and receives messages from *topic1*, and sends alert to the *body temperature sensor* through CIC service. There could be predefined recommendations for the user, for instance, when body temperature of a user is getting high, then user receives an alert to take rest. If there is new data from the device, then *function1* gets an event from Pub/Sub. In addition, Cloud Firestore storage service is used in this use case where function1 sends *body temperature sensor* data in real time controlled Cloud Firestore service for analysis. Cloud IoT application allows to visualize the data for users, such as doctor and nurse, in the hospital and send the notification to doctor, when user's body temperature is higher than normal. Nurses on the other hand have limited access compared to the doctor. Moreover, if there is any emergency or critical condition, the heart rate data is sent to heart specialist for more detailed analysis and recommendations. Currently, the GCP-IoTAC model is based on role-based approach and cannot incorporate the attributes of users. However, we envision that ABAC can enhance the current access control mechanisms in GCP-IoT.

4.5   Limitations and Fine Grained Enhancements

4.5.1   Proposed Enhancements in AWS IoTAC

In AWS IoTAC, attributes for various entities can be utilized to provide more fine grained access control. There are some types of attributes in AWS IoT, however, these attributes do not enable a full ABAC. In ABAC model, attributes of both the users (actors)—requesting access, and attributes of the resources (target objects)— on which accesses are performed, need to be employed and used in the access control policies for enabling fine grained access control on IoT devices. In ABAC, attributes are name-value pairs that represent characteristics of different entities,

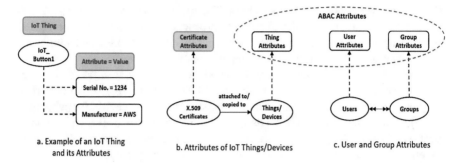

a. Example of an IoT Thing   b. Attributes of IoT Things/Devices   c. User and Group Attributes
   and its Attributes

Fig. 4.14 Attributes in AWS IoT

such as users and objects, and also environment or contextual attributes to make access decisions. Things can have a set of attributes defined in the cloud which are synchronized with their associated physical devices.

Different types of attributes in AWS IoT are shown in Fig. 4.14a. A thing can get attributes through the certificate attachment or association as shown in Fig. 4.14b. While creating a X.509 certificate, a number of attributes are defined, thus, when a certificate is associated with a thing then its certificate attributes can be used in IoT policies for assigning permissions to the thing. However, these certificate attributes do not represent thing properties directly and are different than typical ABAC attributes. Due to the following reasons, AWS IoTAC have a limited form of ABAC model.

- AWS-IoTAC model can only utilize attributes of IoT things/devices that are requesting access on IoT resources in the cloud.
- The thing attributes can be utilized in the policy only when things/devices are using MQTT protocol for connecting and communicating with AWS IoT.
- In AWS IoT, there are limited number of attributes that can be defined for an entity, i.e., only fifty attributes of a thing among which three are searchable attributes.

Therefore, some ABAC enhancements for the AWS-IoTAC model are proposed to incorporate a more complete form of ABAC in AWS-IoTAC.

1. **ABAC with Target Resource Attributes**
   In use case scenario 2, it was shown that AWS-IoTAC model can incorporate attributes of things/devices that perform action(s) in IoT policy for allowing IoT operations on specific things/devices. However, attributes of things/devices on which the operations are being performed (also referred as target attributes) should also be incorporated in AWS-IoTAC model. In addition, attributes of various entities, independent of the connection and communication protocol being used, must be incorporated in the AWS-IoTAC model. The target resource attributes are mainly useful in defining more fine grained access and also allow to isolate the identity of specific IoT objects in policies. For instance, if a user want

to allow an IoT device to publish messages on specific IoT devices with specific attributes and their values. In this case, the device that publishes a message need not be aware of specific topics and devices where it needs to publish messages and rather can publish to multiple topics (of multiple devices) when a set of attributes are satisfied as per the policy.

2. **User and Group Attributes in AWS-IoTAC**

   A more complete form of ABAC need to incorporate attributes of users and groups of users, as depicted in Fig. 4.14c. There are billions of IoT devices that users are using in today's connected world. Therefore, we need to include both users and devices attributes in making access control decisions and enable multiple users to use various IoT devices. Using users, groups, and things/devices attributes in access control policies can enable fine grained access control in cloud-enabled IoT platforms.

3. **The Policy Machine for Policy Management**

   In AWS cloud, a policy-based access control approach is utilized where policy files are attached to entities such as users, groups, "roles", and certificates. For all these entities, there are numerous policies defined and attached to specific entities for enforcing desired permissions. With billions of devices and their users, the access control policies for these entities will scale rapidly becoming unmanageable, thus resulting in a policy-explosion problem similar to the role-explosion. In the future, access control policies need to be managed by customer-based policy management tool for simplifying policy administration. Policy Machine (PM) [28, 29] is an access control policy specification and enforcement tool developed by the National Institute of Standards and Technology (NIST). It is also known as Next Generation Access Control (NGAC) and can be utilized in this context. However, real-world use case implementation utilizing PM for policy specification is required to demonstrate and evaluate its viability.

4.5.2   Proposed Enhancements in GCP IoTAC

This section first highlights limitations of the RBAC model on GCP-IoT, and then presents an attribute-based extension for the current GCP model based on the NIST proposed strategies [30] for combining ABAC and RBAC.

4.5.2.1   Limitations

RBAC is one of the dominant access control models in cloud computing. However, with rapidly evolving cloud and IoT domains, existing RBAC capabilities alone are not sufficient to meet dynamic access control requirements of these domains, especially for E-health, smart homes and cities, smart supply chain management, etc. In these domains, it is a challenge to manage numerous roles for a large number of dynamic users and resources. In GCP access control, users are assigned roles with permissions for accessing cloud resources. There is also a role-hierarchy that enables inherited permissions for multiple users on various resources. However, RBAC has an inherent limitation, i.e., role-explosion problem. In addition, for enabling more fine grained access control in a system, contextual information also need to be considered which is missing in RBAC. Generally, RBAC is static and coarse-grained in nature, and doesn't consider contextual (or environmental) factors such as time or location. This limits the capability of GCP for enabling fine grained access control in dynamic applications.

In Google Cloud IoT, currently there is no mechanism to enable IoT device-level access control. All the access and authorizations are defined on device registries for a user, a group of users, or server-side service account users. Therefore, GCP-IoTAC model lacks a dynamic and flexible access control mechanism due to the limitations of RBAC. To address these limitations, some ABAC extensions that can be applied in GCP IoT are discussed below.

4.5.2.2   Attribute-Based Extensions in GCP

In order to enable fine grained access control in Google cloud, here we present specific approaches for adding attributes to the GCP and its IoT platform. In [30], authors proposed three different strategies to incorporate attributes and roles together. Based on these recommendations, we discuss attribute-based enhancements for the GCPAC model through Dynamic Roles, Attribute-Centric and Role-Centric strategies. These strategies are discussed in the context of GCP object types and objects and their permission assignment.

- **Dynamic Roles:** In [30], dynamic roles approach utilizes User Attributes (UA) for determining the roles of a user. In RBAC, user attributes simplify the user-role assignment (URA) and reduce multitude of role assignments which could result into role-explosion problem. The roles assigned to the user change based on their assigned attributes. This approach could be applied on GCP cloud and IoT access control models by introducing user and group attributes to achieve dynamic roles assignment in the models. Furthermore, the use of attributes can be applied for dynamic permission assignments to roles based on virtual device/device attributes (VDA), registry attributes (a group of virtual devices) and IoT operations.
- **Attribute-Centric:** In attribute-centric approach, a role is a user attribute, unlike roles in RBAC. Here, access decisions are defined based on attributes of entities (user and its group attributes, virtual device/device and its registry attributes). Attribute-centric approach allows to define fine grained authorization in Cloud-Enabled IoT Platforms. Therefore, using this approach, GCP and its IoT access control model can be extended to include User Attribute (UA), Object Types Attributes (OTA), and Object Attributes (OA) which would be used in policies for determining if specific cloud and IoT operations are allowed or not. Attribute-centric approach will completely change current Google Cloud and IoT access

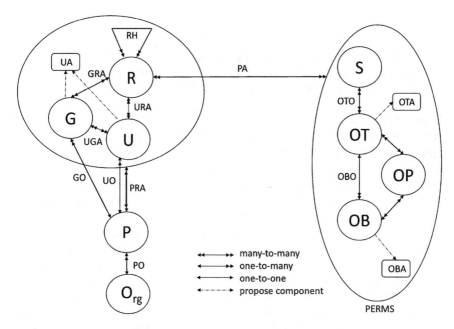

Fig. 4.15 Role-centric approach for GCP access control model

control models by fully converting them to ABAC models. However, it is challenging to drastically change the model without appropriate support and capabilities in specific platforms.

- **Role-Centric:** This approach focuses on assigning maximum permissions through roles, similar to RBAC, and also utilizes attributes of different entities to further restrict some permissions on specific resources. Hence, the overall set of permissions for a user are determined based on its roles and attributes as well as attributes of other entities, as defined in ABAC policies. A role-centric approach to extend Openstack's RBAC model with user attributes has been developed by Bhatt et al. [31]. The authors developed a role-centric ABAC model for OpenStack. A role-centric approach is more intuitive and easily adaptable for GCP cloud and its IoT access control models. Figure 4.15 shows the proposed role-centric approach in the GCPAC model where the final set of permissions are determined based on UA, OTA, and OA after the roles with permissions have been assigned. In the current GCPAC model, Cloud IAM conditions allow to define fine grained access control on object types and objects by specific users, only if predefined conditions in the policy are met. However, this is not available for GCP-IoT service at this time. Moreover, a similar approach can also be easily adapted for GCP-IoTAC model.

While the above approaches are discussed specifically for GCPAC, these approaches can similarly be adapted to enhance the GCP-IoTAC model with ABAC capabilities.

4.6 Summary

In this chapter, we discussed formal access control models for two of the leading cloud service providers, AWS and GCP. First, we presented formal access control models for AWS cloud platform, known as AWSAC, and its IoT service, known as AWS-IoTAC model. Second, we present access control models for Google Cloud platform (GCP), named as GCPAC model, and GCP IoT access control (GCP-IoTAC) model. use cases from two different IoT domains, smart home and E-Health, are demonstrated for AWS cloud-IoT platform and GCP cloud-IoT platform respectively. There are a large number of cloud and IoT services provided by AWS and GCP, and its a challenge to capture all of these capabilities while developing the models and use cases. However, we mainly focused on access control aspects of these platforms and their capabilities to define fine grained authorizations for various entities in specific platforms. These models are represented as a blueprint for developing more advanced access control models for future Cloud-Enabled IoT architectures. Finally, some limitations of these models were discussed with proposed ABAC enhancements for each cloud-IoT platform, AWS and GCP. Based on real-world use cases and ABAC enhancement strategies, role-centric approach seems promising for enabling fine grained access control in GCP. However, more research on these strategies in the context of cloud-enabled IoT access control is needed.

References

1. Zhang, Y., Patwa, F., & Sandhu, R. (2015).   Community-based secure information and resource sharing in AWS public cloud. In *1st IEEE Conference on Collaboration and Internet Computing (CIC)* (pp. 46–53). IEEE.
2. AWS IoT Platform.   https://docs.aws.amazon.com/iot/latest/developerguide/what-is-aws-iot. html. Accessed: 2017-01-08.
3. Ouaddah, A., Abou Elkalam, A., & Ouahman, A. A. (2017a).   Towards a novel privacy-preserving access control model based on blockchain technology in IoT.   In *Europe and MENA Cooperation Advances in Information and Communication Technologies* (pp. 523–533). Springer.
4. Anggorojati, B., Mahalle, P. N., Prasad, N. R., & Prasad, R. (2012). Capability-based access control delegation model on the federated IoT network. In *The 15th International Symposium on Wireless Personal Multimedia Communications* (pp. 604–608). IEEE.
5. Ameer, S., & Sandhu, R. (2021).   The habac model for smart home IoT and comparison to egrbac.
6. Zhang, G., & Tian, J. (2010).   An extended role based access control model for the internet of things. In *2010 International Conference on Information, Networking and Automation (ICINA)* (vol. 1, pp. V1–319). IEEE.
7. Ameer, S., Benson, J., & Sandhu, R. (2020).   The egrbac model for smart home IoT. In *2020 IEEE 21st International Conference on Information Reuse and Integration for Data Science (IRI)* (pp. 457–462). IEEE.
8. Gupta, D., Bhatt, S., Gupta, M., Kayode, O., & Tosun, A. S. (2020a).   Access control model for google cloud IoT.   In *2020 IEEE 6th Intl Conference on Big Data Security on Cloud*

(BigDataSecurity), IEEE Intl Conference on High Performance and Smart Computing,(HPSC) and IEEE Intl Conference on Intelligent Data and Security (IDS) (pp. 198–208). IEEE.

9. Gupta, M., Awaysheh, F. M., Benson, J., Al Azab, M., Patwa, F., & Sandhu, R. (2020b). An attribute-based access control for cloud-enabled industrial smart vehicles. *IEEE Transactions on Industrial Informatics*. https://doi.org/10.1109/TII.2020.3022759.

10. Alshehri, A., & Sandhu, R. (2017). Access control models for virtual object communication in cloud-enabled IoT. In *2017 IEEE International Conference on Information Reuse and Integration (IRI)* (pp. 16–25). IEEE.

11. Gupta, M., Benson, J., Patwa, F., & Sandhu, R. (2020c). Secure V2V and V2I communication in intelligent transportation using cloudlets. *IEEE Transactions on Services Computing*. https://doi.org/10.1109/TSC.2020.3025993.

12. Gupta, M., Benson, J., Patwa, F., & Sandhu, R. (2019). Dynamic groups and attribute-based access control for next-generation smart cars. In *Proceedings of the Ninth ACM Conference on Data and Application Security and Privacy* (pp. 61–72).

13. Ouaddah, A., Mousannif, H., Elkalam, A. A., & Ouahman, A. A. (2017b). Access control in the Internet of Things: Big challenges and new opportunities. *Computer Networks, 112*, 237–262.

14. Amazon Web Services (AWS). https://aws.amazon.com/. Accessed: 2016-12-10.

15. Microsoft Azure. https://azure.microsoft.com/en-us/. Accessed: 2016-11-28.

16. Google Cloud Platform. https://cloud.google.com/. Accessed: 2016-12-10.

17. Sandhu, R., Coyne, E. J., Feinstein, H., & Youman, C. (1996). Role-based access control models. *Computer, 29*(2), 38–47.

18. Ferraiolo, D. F., Sandhu, R., Gavrila, S., Kuhn, D. R., & Chandramouli, R. (2001). Proposed NIST standard for role-based access control. *ACM Transactions on Information and System Security (TISSEC), 4*(3), 224–274.

19. Overview of Internet of Things. https://cloud.google.com/solutions/iot-overview/. Accessed: 2016-12-10.

20. Azure IoT. https://docs.microsoft.com/en-us/azure/iot-hub/iot-hub-what-is-iot-hub. Accessed: 2016-11-10.

21. X.509 Certificates. https://searchsecurity.techtarget.com/definition/X509-certificate. Accessed: 2017-02-10.

22. MQTT.fx—A JavaFX based MQTT Client. https://www.mqttfx.org/. Accessed: 2016-09-10.

23. Nitti, M., Pilloni, V., Colistra, G., & Atzori, L. (2016). The virtual object as a major element of the Internet of Things: A survey. *IEEE Communications Surveys & Tutorials, 18*(2), 1228–1240.

24. Bhatt, S., Patwa, F., & Sandhu, R. (2017). Access control model for aws internet of things. In *International Conference on Network and System Security* (pp. 721–736). Springer.

25. Alshehri, A., & Sandhu, R. (2016). Access control models for cloud-enabled Internet of Things: A proposed architecture and research agenda. In *2nd IEEE International Conference on Collaboration and Internet Computing (CIC)* (pp. 530–538). IEEE.

26. AWS SDK for JavaScript in Node.js. https://aws.amazon.com/sdk-for-node-js/. Accessed: 2016-08-10.

27. Google Cloud Platform. https://cloud.google.com/docs. [Online; accessed 10-February-2020].

28. Ferraiolo, D., Gavrila, S., & Jansen, W. (2014). Policy Machine: Features, architecture, and specification. *NIST Internal Report 7987*.

29. Ferraiolo, D., Atluri, V., & Gavrila, S. (2011). The Policy Machine: A novel architecture and framework for access control policy specification and enforcement. *Journal of Systems Architecture, 57*(4), 412–424.

30. Kuhn, D. R., Coyne, E. J., & Weil, T. R. (2010). Adding attributes to role-based access control. *Computer, 43*(6), 79–81.

31. Bhatt, S., Patwa, F., & Sandhu, R. (2016). An attribute-based access control extension for OpenStack and its enforcement utilizing the Policy Machine. In *2nd IEEE International Conference on Collaboration and Internet Computing (CIC)* (pp. 37–45). IEEE.

32. Cathey, G., Benson, J., Gupta, M., & Sandhu, R. (2021). Edge centric secure data sharing with digital twins in smart ecosystems. *Preprint arXiv:2110.04691.*
33. Bhatt, S., Pham, T. K., Gupta, M., Benson, J., Park, J., & Sandhu, R. (2021). Attribute-based access control for AWS internet of things and secure Industries of the Future. *IEEE Access, 9,* 107200–107223.
34. Gupta, M., & Sandhu, R. (2021). Towards activity-centric access control for smart collaborative ecosystems. In *Proceedings of the 26th ACM Symposium on Access Control Models and Technologies* (pp. 155–164).
35. Awaysheh, F. M., Alazab, M., Gupta, M., Pena, T. F., & Cabaleiro, J. C. (2020). Next-generation big data federation access control: A reference model. *Future Generation Computer Systems, 108,* 726–741.
36. Gupta, M., & Sandhu, R. (2016). The GURAG administrative model for user and group attribute assignment. In *International Conference on Network and System Security* (pp. 318–332). Springer.
37. Gupta, M., & Sandhu, R. (2018). Authorization framework for secure cloud assisted connected cars and vehicular internet of things. In *Proceedings of the 23rd ACM on Symposium on Access Control Models and Technologies* (pp. 193–204).
38. Gupta, M., Patwa, F., & Sandhu, R. (2018). An attribute-based access control model for secure big data processing in hadoop ecosystem. In *Proceedings of the third ACM workshop on attribute-based access control* (pp. 13–24).

Chapter 5
Secure Virtual Objects Communication

5.1 Introduction

The Internet of Things (IoT) technology is invented from the development of wireless communication systems. Technologies such as sensing, networking, software architectures, data analytics, information management, and visualization all meet in the IoT. However, there are some security issues that have accompanied with the emergence of the IoT technology that calls for a major review of existing security solutions involving access control systems. Various access control models for IoT have been recommended to handle security and privacy concerns [1–3].

In Chap. 2, the access-control oriented architecture (ACO) [4] is discussed, which is for cloud-enabled IoT including four layers: an object layer, a virtual object (VO) layer, a cloud services layer, and an application layer. Also, the benefit of using virtual objects in the IoT [5] is presented. ACO identifies the need to control communication within each layer and across adjacent layers. Also, it recognizes the need to control access to data in cloud services and application layers. In this chapter, the focus will be on proposing access control models for VO communications, within the context of the developed ACO framework. Also, it will be shown that it is possible to use AWS IoT mechanisms to effectively demonstrate and control VO communications using the services. Virtual objects in the IoT can communicate in different ways. The most common method used in VO communication style is topic-based publish-subscribe (see Chap. 1, Sect. 1.3). In Amazon Web Service (AWS) IoT, for example, a virtual object is called a shadow that uses reserved MQTT topics to allow applications and things to communicate with shadows and operate operation such as get, update, or delete the state information for a device [6–8].

Access control models like access control lists (ACLs), capability lists, and role-based access control (RBAC) are traditional models that are used as operational models and administrative models [9]. Also, a broader model that encompasses the benefits of the three traditional models is Attribute-based access control (ABAC) [10, 11] that has lately received attention, and can also be applied in IoT for

M. Gupta et al., *Access Control Models and Architectures For IoT and Cyber Physical Systems*, https://doi.org/10.1007/978-3-030-81089-4_5

fine grained authorization policies. In addition, ABAC introduces new features appropriate for dynamic and open environments of IoT. It is vital to control VO communication by means of access control mechanisms. It could also be useful to use multiple access control mechanisms in this regard [12]. In this chapter, the virtual objects communication style in [13, 22] is presented as well as the access control models for VO communication in two layers: operational models and administrative models, assuming topic-based publish-subscribe interaction method among VOs. Operational models for VO communication are presented in ACLs for topics and capabilities for virtual objects as well as ABAC operational model. Administrative models for the two operational models are presented using (i) ACLs, (ii) RBAC and (iii) ABAC.

AWS IoT is a large commercial cloud-IoT platform that confirms its suitability to the above-mentioned academic models for controlling ACO and VO communications. Though AWS IoT has the concept of digital shadows that is mainly corresponding to VOs, it lacks the clear method of VO communication and therefore no VO communication control. In this chapter, the method of VO communication model in the AWS IoT is clarified, and discuss the access control model for virtual objects (shadows) communication in AWS IoT is called AWS-IoT-ACMVO [14]. In the later part of this chapter, we will present a proof-of-concept implementation of the two speeding cars use cases in AWS IoT under structure of this IoT-ACMVO model, and it will present selected performance measurements. This entire chapter will demonstrate on one side the formal academic models developed (independent of AWS IoT) can be enforced in the notion of digital twins. It further proposes extensions to AWS IoT that would be important to facilitate such implementation.

5.1.1   Chapter Organization

This chapter will have the the foundational access control models and proposed implementation in AWS IoT for virtual object communication. Section 5.2 will discuss the developed operational access control models for virtual object communication, which will propose the ACL and Capability based solution followed by ABAC supported solution. Section 5.3 will elaborate on the administrative models to support the operational access control models for VO, proposing the ACL, RBAC and ABAC based administrative models. The following sections will focus on the implementation in AWS IoT. Section 5.4 will present an access control model for virtual object communication referred to as AWS-IoT-ACMVO. A discussion of some issues of AWS IoT and possible enhancements are explained in Sect. 5.5. Section 5.6 will discuss the use case of ACO-IoT-ACMsVO within the AWS-IoT-ACMVO model, and elaborate the proof-of-concept implementations of the use case in two scenarios in AWS IoT platform. At the end, the performance aspects of the implementation is also highlighted.

5.2 Operational Access Control for VO Communication

The access control models for VO communication is presented in two layers: the operational models and the administrative models. The operational models highlight which VOs are authorized to communicate with other VOs via MQTT topics. On the other hand, the administrative models specify how to control the operational models. The idea of separating the operational and administrative models is first developed in RBAC model where operational models were defined in [15, 16, 23–26] and administrative models in [11, 17].

In this section, two operational models are developed which is ACLs and capabilities-based access control, and attribute-based access control. Publish/subscribe schemes basically engage Message Brokers (MBs) [18] to direct messages from publisher to subscribers for topics. Subscribers first need to register (by sending a subscribe request) with a message broker of a topic, then published messages that is sent by publisher to message broker will be forwarded to all subscribers. The authorizations policy of this communication will control VO subscriptions and publishing, where only the authorized VOs are allowed to publish or subscribe to a topic. This type of authorization regulates the allowed pattern of communication in the VO layer, and thus in indirect way the object layer. Some questions can be addressed by operational access control models. Which VOs are allowed to publish or send a subscription request to a topic's MB? Where should topics MB forward data to? Which MBs need VOs publish to or transmit a subscription request to? Which MBs should VOs get data from? These result the following linked question: Where must the publish and subscribe controls be placed? Should be placed in the virtual object side, the topic side, or both?

In the operational models there are two recognized sets of entities: Topics (T) and Virtual Objects (VO), and a set of Rights R={p,s}, where p means publish and s means subscribe. VOs are considered as active entities because they can publish/receive data to/from subscribed topics. MB will be associated to each topic in order to reply to subscribe requests from VOs, get published data from a VO to a topic, and forwards received data to the subscribers of the topic. The forward operation will be represented in the singleton set F={Forward}. The nature of the traditional resource/object and user/subject entities in access control models [10, 16, 19–21, 27, 28] is different from the entities of the operational models.

5.2.1 ACL and Capability Based (ACL-Cap) Operational Model

In this section, the ACL-Cap model is discussed, which merges capability lists (Cap) for VOs and ACLs for topics as explained in Fig. 5.1. Administrators will maintain these lists as will be described in administrative access control models section. In the topic side, it shows the ACL of a topic that contains a list of VOs as well as a publish or subscribe right for each VO. Similarly, in the VO side, it shows the

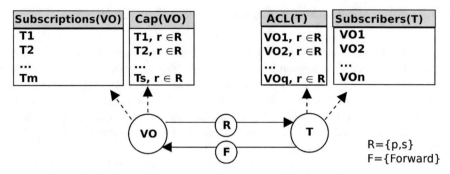

Fig. 5.1 The ACL-Cap model

capability list of a VO that contains a list of topics with the publish or subscribe right for each topic. The capability list authorizes VO in the list to publish or subscribe to a topic. Likewise, the ACL of a topic informs the topic's MB which VOs can subscribe or publish to it. Because the VOs and topic MBs are fully automated, this dual ACL-Cap approach is more convenient and secure relative to ACL-only or capability-only approaches. The ACL-Cap scheme deny unauthorized operations as soon as possible, rather than postponing decisions until later.

A certain VO that has a publish capability is authorized to publish to a topic. The VO will successfully publish only if that topic's ACL has an analogues entry for that VO with the publish right. The publish authorization rule is thus expressed as follows.

$$Auth\text{-}Publish(VO, T) \equiv (T, p) \in Cap(VO) \wedge$$
$$(VO, p) \in ACL(T) \tag{5.1}$$

The more complicated operation is subscribing. Because the subscribe relationship require to be initiated prior published data is forwarded and received. Thus, a request to subscribe from a VO to a topic is required as well as an accepting response from the topic's MB. This is a multi-step operation, the authorization rule for successful completion of subscribe is expressed as follows.

$$Auth\text{-}Subscribe(VO, T) \equiv (T, s) \in Cap(VO) \wedge$$
$$(VO, s) \in ACL(T) \tag{5.2}$$

A full subscribe operation scenario is adding the topic T to the VO's subscriber list, and the VO to the topic's subscriber list as shown in Fig. 5.1. A *Forwarding* operation of published data by a topic's MB to a VO is expressed as follows.

$$Auth\text{-}Forward(T, VO) \equiv VO \in Subscribers(T) \wedge$$
$$T \in Subscriptions(VO) \tag{5.3}$$

Table 5.1 ACL of topics

$T1$	....	$Tn\text{-}1$	Tn
$VS1, \text{p}$	....	$VSn\text{-}1, \text{p}$	VSn, p
$VS2, \text{s}$	....	VSn, s	$VC1, \text{s}$

Table 5.2 Capability list of VOs

$VS1$	....	VSm	$VC1$
$T1, \text{p}$	....	Tn, p	Tn, s
	....	$Tn\text{-}1, \text{s}$	

Equations 5.1 and 5.2 respectively answer some questions: which VOs are authorized to send a subscription request or publish to a topic's MB? Equation 5.3 answers the question as to which VOs a topic's MB can forward data to. Note that Eq. 5.1 can be partially checked at the publishing VO's side, thus banning an evil VO from deliberately or unintentionally publishing to unauthorized topics (as would be possible in an ACL-only approach). Tables 5.1 and 5.2 show the $T = \{T1, .., Tn - 1, Tn\}$ and $VO = \{VS1, .., VSn, VC1\}$ with the ACLs of topics and capability lists of virtual objects.

5.2.2   ABAC Operational Model

This section shows the ABAC operational model which is explained in Fig. 5.2. The ABAC operational model entities are similar to ACL-Cap operational model, which are the set VO of virtual objects and the set T of topics with rights R={p,s} and F={Forward}. In addition, there are a set of attributes, TA for topic attributes, and VOA for virtual object attributes as follows.

$$VOA = \{VO\text{-}Publish, VO\text{-}Subscribe, VO\text{-}Subscriptions,$$

$$VO\text{-}Location\}$$

$$TA = \{T\text{-}Publish, T\text{-}Subscribe, T\text{-}Subscribers,$$

$$T\text{-}Location\}$$

Attributes like the VO-location and T-location are atomic valued. The location attribute provides the location of the linked topic or VO in suitable units. The rest of attributes are set-valued. A list of values for VO-Publish, VO-Subscribe, and VO-Subscriptions are a subset of the topics T. Values for T-Publish, T-Subscribe, and T-Subscribers are a subset of the virtual objects VO. The following authorization rules express the same policy as in Sect. 5.2.1.

Fig. 5.2 ABAC operational model

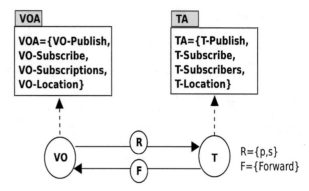

$$Auth\text{-}Publish(VO, T) \equiv T \in VO\text{-}Publish(VO) \wedge$$
$$VO \in T\text{-}Publish(T) \tag{5.4}$$

$$Auth\text{-}Subscribe(VO, T) \equiv T \in VO\text{-}Subscribe(VO) \wedge$$
$$VO \in T\text{-}Subscribe(T) \tag{5.5}$$

$$Auth\text{-}Forward(T, VO) \equiv T \in VO\text{-}Subscriptions(VO)$$
$$\wedge VO \in T\text{-}Subscribers(T) \tag{5.6}$$

The attributes VO-Publish, VO-Subscribe, T-Publish, and T-Subscribe are assigned by administrators. The VO-Subscriptions and T-Subscribers attributes are assigned as a consequence of establishing the subscribe relationship as discussed in Sect. 5.2.1. ABAC operational model has additional improvements that ACL-Cap operational model doesn't have which are the T-location and VO-location attributes.

It is assumed that VO-Location, which is assigned by an administrator, is an attribute that is automatically assigned to be the location received from the physical sensor. The following condition can be added to each of the above equations which are in Eqs. 5.4, 5.5, and 5.6.

$$\text{-}Location(T) \approx VO\text{-}Location(VO) \tag{5.7}$$

Adding the equation will further restrict the communication pattern amongst the VOs by taking their location into consideration. In particular, if sensors are moved to elsewhere, then the granted authorized communication will be revoked. Thus, movements of VOs will be accommodated because of the condition of location matching. A single ABAC authorization rule integrates virtual object and topic attributes. In this regard Eqs. 5.4, 5.5 and 5.6, are respectively comparable to Eqs. 5.1, 5.2 and 5.3. ABAC operational model, however, has additional attributes such as in Eq. 5.7 that the ACL-Cap operational model doesn't have.

5.2.3 RBAC Limitations

This section argues some RBAC restrictions in context of IoT VO communications. RBAC is proposed with the concept of assigning users to roles through which users primarily obtain permissions to execute operations on targeted objects. However, topics and virtual objects do not match this model very precisely. Virtual objects are active entity that interacting with other entities via publish and subscribe operations, and receiving input via forward operations. Likewise, topics are targets to publish operation, and actors in terms of redirecting and accepting subscription requests. The active aspects of virtual objects and topics could be adjusted in RBAC by giving these entities to alternately special sets of roles. Looking at the Eqs. 5.1 and 5.2, the first piece of the equations (i.e., $(T, p) \in Cap(VO)$ and $(T, s) \in Cap(VO)$) could be illustrated in RBAC operation model by permission assignment of topic permissions to the VO's role. The second piece (i.e., $(VO, p) \in ACL(T)$ and $(VO, s) \in ACL(T)$) could likewise be illustrated by permission assignment of virtual objects permissions to T's role. But this separates the equations into discrete roles, which should thereby both be taken into account when access assessments are made. This consideration of roles of both actor and target needs core addition to conventional RBAC [16].

5.3 Administrative Access Control for VO Communication

This section shows three administrative access control models to regulate VO communication by employing ACL, RBAC and ABAC methods. An administrative model is an important supplement to the operational models explained earlier. Also, the construction of an administrative model is not firmly attached with that of the operational model, as will manifest.

The expression admins is used to denote users who are entitled to regulate VO communication, by adjusting setting of the operational model. For simplicity's sake, it is assumed that topics admins are also VOs admins. The operational ACL-Cap model is controlled by admin through two core questions: Who is authorized to add or delete (T, p) or (T, s) from Capability list of VO? Who is authorized to add or delete (VO, p) or (VO, s) from ACL of T? On the other hand, the operational ABAC model is controlled by admin through two core questions: Who is authorized to designate or eliminate values to/from attributes of T? Who is authorized to designate or eliminate values to/from attributes of VO?

Fig. 5.3 Administrative ACL

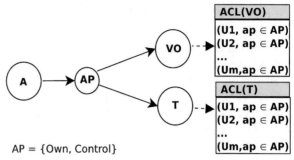

AP = {Own, Control}

5.3.1 Administrative ACL Model

In addition to the operational model mentioned, the administrative ACL model presents a set of admin users (A) and admin permissions (AP) as follows.

$$A = \{U1, .., Um\text{-}1, Um\}$$

$$AP = \{Own, Control\}$$

For each T and VO the administrative ACL model has one ACL as presented in Fig. 5.3. The two permissions: *Own* and *Control* are alike in terms of allowing modifications to ACLs, Capability lists, and administered attributes of topics and virtual objects. The difference between the Own and Control permissions is that Own permits the admin user to grant Own or Control permission over the virtual object or topic to other admin users, whereas Control permission does not.

If the pair (U, ap) is respectively in the ACL of T or VO, where ap is Own or Control permission, then certain admin user U can control T or VO. The authorization rule for U to control T or VO can be expressed as follow.

$$Auth\text{-}Control(U, T) \equiv (U, ap) \in ACL(T) \tag{5.8}$$

$$Auth\text{-}Control(U, VO) \equiv (U, ap) \in ACL(VO) \tag{5.9}$$

5.3.1.1 Administrative ACL Model for Operational ACL-Cap

The administrative ACL model for operational ACL-Cap model is represented in two different configurations in Tables 5.3 and 5.4. Table 5.3 shows that all admin users possess the Own permission for all topics and virtual objects, while in Table 5.4 only user1 (U1) does. Apparently U1 has granted U2 and U3 the control of virtual sensors VS1 to VSn, and topics T1 to Tn-1. Also, U1 has granted control over VC1 to admin U4.

Table 5.3 All admins have own permission for all VO and T

T1, VS1 Admins	T2, VS2 Admins	.....	Tn, VSn Admins	VC1 Admins
(U1, Own)	(U1, Own)	.....	(U1, Own)	(U1, Own)
....	....	.....	....	....
(Um, Own)	(Um, Own)	.....	(Um, Own)	(Um, Own)

Table 5.4 Only user U1 has own permission

T1, VS1 Admins	T2, VS2 Admins	.....	Tn, VSn Admins	VC1 Admins
(U1, Own)	(U1, Own)	.....	(U1, Own)	(U1, Own)
(U2, Control)	(U2, Control)	.....	(U2, Control)	(U4, Control)
(U3, Control)	(U3, Control)	.....	(U3, Control)	

5.3.1.2   Administrative ACL Model for Operational ABAC

The structure of administrative ACL model does not change, but the meaning of
Own and Control are altered to the ABAC operational model. The differentiation
between Own permission and Control permission last as mentioned overhead, and
only effects the administrative ACLs. However, in operational ABAC the Control
permission over a topic or virtual object allows the admin to correspondingly
modify topic or virtual object attributes, which are administrable. These are T-
Publish, VO-Publish, VO-Subscribe, T-Subscribe and T-Location in our use case.
VO-Subscriptions, VO-Location and T-Subscribers are automatically allocated and
not managed by admins.

The administrative ACL model has one ACL for each topic and virtual object in
both cases overhead. Hence, with large numbers of virtual objects and topics this
model will be very difficult to manage.

5.3.2   Administrative RBAC Model

The administrative RBAC model remains to use the set of admin permissions $AP = \{Own, Control\}$, and admin users $A = \{U1, .., Um\text{-}1, Um\}$, that are presented
in Sect. 5.3.1. Moreover, it proposes a set of administrative roles (AR) and admin
permissions (AP) as follows.

$$AR = \{AR1, .., ARs\}$$

$$AP = (VO \times AP) \cup (T \times AP)$$

If a user (U) has admin assignment (AA) with one administrative role AR_1, and
AR_1 coupled through admin permission assignment (APA) with a virtual object or

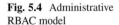

Fig. 5.4 Administrative
RBAC model

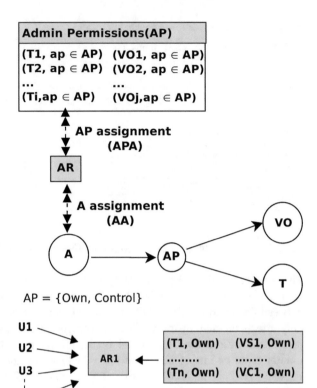

$AP = \{Own, Control\}$

Fig. 5.5 Administrative
RBAC: Reflects Table 5.3

a topic, then a user U can control the topic or the virtual object. The authorization
scenario is presented in Fig. 5.4.

The administrative RBAC model is simpler to preserve than administrative ACL,
because of the well-known benefits of RBAC over per-topic and per-VO ACLs. The
number of administrative roles that request to be maintained is decreased to one in
the case of Table 5.3 as indicated in Fig. 5.5, and to three for the case of Table 5.4
as indicated in Fig. 5.6. These are fixed numbers in contrast to the linear growth in
ACLs with an increase in hypothetical subjects and objects.

5.3.3 Administrative ABAC Model

The administrative ABAC model for our use case remains to use the set of admin
permissions $AP = \{Own, Control\}$ and admin users $A = \{U1, .., Um\text{-}1, Um\}$,
presented in Sect. 5.3.1. In addition, it introduces administrative attributes for VOs
(VOAA), topics (TAA), and users (UAA), as follows.

Fig. 5.6 Administrative
RBAC: Reflects Table 5.4

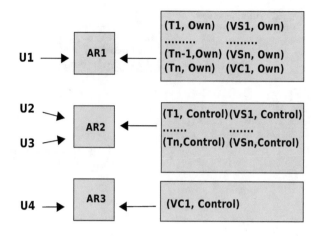

$$TAA = \{T\text{-}Location, T\text{-}Department\}$$

$$VOAA = \{VO\text{-}Type, VO\text{-}Location, VO\text{-}Department\}$$

$$UAA = \{U\text{-}Type, U\text{-}Location, U\text{-}Department\}$$

These administrative attributes are basically reusing the operational attributes, which are mentioned in the operational ABAC model in Sect. 5.2.2 for Location, as well as adding additional administrative attributes which are Type and Department. Administrative attributes are set to be atomic valued. In both Type and Department attributes, the range will be some small number of counted items. Figure 5.7 illustrates TAA, VOAA, and UAA actuality used to grant admin permission (AP) for A. The explanation of the Own and Control permissions for both operational models is discussed in Sect. 5.3.1.

The Control permission can be granted to users over objects or topics through the following stated authorization rules:

$$Auth\text{-}Control(U, VO) \equiv$$

$$(U\text{-}Type(U) = Own \vee U\text{-}Type(U) = Control) \wedge$$

$$U\text{-}Department(U) = VO\text{-}Department(VO) \wedge$$

$$(VO\text{-}type = sensor \vee VO\text{-}type = camera) \wedge$$

$$U\text{-}location \approx VO\text{-}Location(VO)$$

$$Auth\text{-}Control(U, T) \equiv$$

$$(U\text{-}Type(U) = Own \vee U\text{-}Type(U) = Control) \wedge$$

$$U\text{-}Department(U) = T\text{-}Department(T) \wedge$$

$$U\text{-}location = T\text{-}Location(T)$$

Fig. 5.7 Administrative ABAC

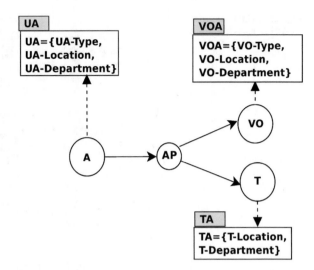

These defined equations grant the *Control* permission to a user over a VO only if users have Own or Control type, users and VOs in same Department and approximate Location, and VO type is sensor or camera. Also, the *Control* permission is granted to a user over a T only if users have Own or Control type, and users and VOs are in same Department and exact Location (remember topic-location attribute is an administered attribute).

In ABAC these rules can be simply altered or refined, e.g., we could have two rules for cameras and sensors. ABAC summaries identity, role, and resources information of ACL and RBAC methods into VO, topic and user attributes. Thus, we can say that ABAC is flexible, scalable and adaptable. In addition, gathered data (e.g. VO-Location) can be extracted to be used as values of attributes, that cooperate with other attributes for decision making.

5.4   AWS-IoT-ACMVO Model for AWS IoT Shadows Communication

This section discusses an access control model for virtual objects (shadows) communication called AWS-IoT-ACMVO as an abstracted view of AWS IoT capabilities. Figure 5.8 illustrates the main components of this model, viz., certificates, policies, MQTT topics, virtual objects (device shadows), and rule engine and its action. The functionalities of components are discussed below.

For devices authentication, AWS IoT utilizes X.509 certificates as the identity credential [7]. Certificates can be either an AWS IoT generated certificate or a certificate signed by a AWS IoT registered external certification authority. In general, one certificate can be provided to different **devices**, but it is suggested

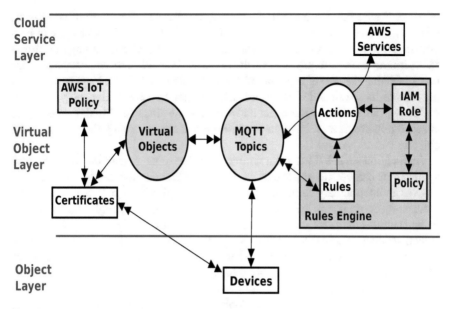

Fig. 5.8 The AWS-IoT-ACMVO components

that each device has a unique certificate to enable fine grained device management. Figure 5.8 displays that every certificate can be provided to one device or more, and multiple certificates can be given to a one device. However, whenever a device connects, only one certificate can be activated.

Whenever a certificate is created, there will be two AWS IoT entities that require to be linked to the certificate to authorize and authenticate AWS IoT devices that need to connect with virtual objects (device shadows), viz., **AWS IoT policy** and **virtual objects**. An AWS IoT policy is a JSON document that is linked to a certificate for the purpose of authorization. It includes one or more policy statements, each policy statement states effect, action, resources, and optional condition. An action is a procedure that can be given or denied to a resource as defined by the value of effect. Actions can be either thing shadow policy actions or MQTT policy actions. The MQTT policy actions are the processes that handle connecting, receiving, or sending data, which are iot:Connect, iot:Subscribe, iot:Publish, and iot:Receive. On the other side, thing shadow policy actions handle permissions in order to deal with thing shadow entity, which are iot:DeleteThingShadow, iot:GetThingShadow, and iot:UpdateThingShadow. Figure 5.8 express that more than one certificate can be linked to each AWS IoT policy, and multiple AWS IoT policies can be attached to each certificate. Thus, the AWS IoT policy is linked to a certificate to permit actions (permissions) to devices that hold the linked certificate (and its private key).

Virtual objects (device shadows) moreover need to be linked to a certificate as a reference that the device is entirely or partially permitted to access. A device shadow can be given one certificate or more, and a one certificate can be linked to one device or more. Figure 5.8 displays relationship between certificates and virtual objects, which is the many-to-many relationship. In AWS-IoT a JSON document can be a virtual object that stores information about a connected device regarding the current state and the desired future state of the connected device. The device shadow has many benefits like the information of device shadow could be utilized to get or set the state of its physical device, even if they are not connected. Generally, a device will have privileges to interconnect and access to the attached virtual objects on the attached policies, if it is linked to a certificate that has attached policies and virtual objects.

Applications in AWS IoT cannot directly update or get data of devices. The way applications can update or get data of a device is through the device shadows that work as an intermediary communication between applications and physical devices. Virtual object can interact with applications or devices only through **MQTT topics**. In another meaning, by publishing or subscribing to MQTT topics of a virtual object, published/subscribed applications and devices are allowed to get, update, or delete the state information of the virtual object (device shadow). The path of MQTT topics start with $aws/things/thingName/shadow/$#, where thingName is a virtual object name, and # is thingName topics that can be utlized to interact with the thingName. For each virtual object there are reserved MQTT topics, that can be employ to publish or subscribe to the virtual object. To publish data to a thing shadow, the following virtual object topics can be used: /update, /get, or /delete. However, when a virtual object wants to publish an acknowledgement about accepting or rejecting the published (received) data, the following topics can be used: /update/accepted, /update/reject, /update/delta, /update/documents, /get/rejected, /get/accepted, /delete/accepted, and /delete/rejected topics. In general, once a virtual object is generated, AWS IoT service automatically creates reserved MQTT topics for the virtual object, and they will be used to publish or subscribe to the generated virtual object. Every virtual object will have certain reserved MQTT topics, and every reserved MQTT topic is related to only one virtual object as shown in Fig. 5.8. Furthermore, if a device has an authorized certificate, then the device can publish or subscribe to one or more MQTT topic. Similarly, if devices have authorized certificates, then each MQTT topic can be used by many authorized devices.

A valuable technique in AWS IoT is that a message published to an MQTT topic can be identified and evaluated by a rule. **Rules** offer processing for the delivered messages to MQTT topics and allow interactions with many AWS services. A rule contains a rule name, SQL statement, optional description, SQL version, and one or more actions. The SQL statement can be used to filter arrived (published) messages to MQTT topics, and next the rule engine redirects filtered messages to AWS services or republishes them to other MQTT topics by utilizing the action field designated in the rule. There are unchanging AWS actions that can be chosen, such as adding a message into a DynamoDB table, invoking a Lambda function, and republishing messages to AWS IoT topics. Thus, rules that are linked to MQTT topics obviously offer a method for virtual objects to interact with AWS services or republish the arrived messages to other MQTT topics (reserved or unreserved).

Every rule can be triggered by more than one topic, and each topic can trigger more than one rule as shown in Fig. 5.8. Moreover, many actions can be performed when a rule is triggered.

When rules redirect the received messages to another AWS service like AWS Lambda, AWS identity and access management **(IAM) role** is controlling the access authorization of other services and the actions of other services. Every IAM role is attached with minimum one **policy** that offers permissions to approach resources stated in the action of the rule or to control actions toward the arrived data. For instance, if an Amazon SNS rule is generated, an IAM role will be linked to the SNS rule to permit access to SNS properties. The linked role have policies that permit actions, such as sns:Publish, toward certain properties in Amazon SNS. Likewise, an IAM role will be linked to the lambda function if a lambda rule is generated. The linked IAM role will have a policies that permit actions, such as iot:Publish and iot:GetThingShadow, toward certain properties in AWS Lambda. Therefore, the IAM role and its linked policies are a piece of the AWS IoT rule definition to control actions. Every action of a rule can only be linked to one IAM role, but every IAM role can be used by many rule actions as shown in Fig. 5.8. Additionally, one IAM role could be linked to many policies, and one policy could be linked to many IAM roles.

5.5 Issues in Enforcing ACO-IoT-ACMsVO Within AWS-IoT-ACMVO

AWS IoT does not provide straight communication among VOs, because VOs are only permitted to communicate through its reserved topics. The AWS-IoT-ACMVO model is one method to provide VO communication through rules within AWS IoT. AWS-IoT-ACMVO maintains the transient data within the virtual object layer without persistent storage while only data is propagated to the higher layers. Consequently, the data privacy is preserved. Figure 5.8 shows all sections of VO communication that participate in the interaction between VO. The ACO-IoT-ACMsVO academic model presumed the communication regime shown in Fig. 5.9 where the communication means between two VOs is a shared topic that VO1 publishes and VO2 subscribes. The rules engine of AWS IoT serves as a communication channel between VOs in the AWS-IoT-ACMVO model as shown in Fig. 5.10. The rules engine allows an analogous effect to be accomplished in AWS IoT.

The control points in ACO-IoT-ACMsVO are located on both VO side and topic (T) side [13] to permit VO to VO communication via topics. For instance, in case of ACL-Cap operational model, a VO can be authorized to publish to a topic when the VO is within the access control list of T and the topic is within the capability list of the VO. In case of the ABAC operational model, a publish right will be permitted if the VO is within T-Publish attribute values and the topic is within VO-Publish attribute values. Similarly, the subscription will be authorized on both topic and VO.

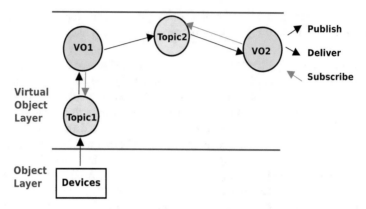

Fig. 5.9 The publish/subscribe topic-based scheme in the ACVO-ACO-IoT

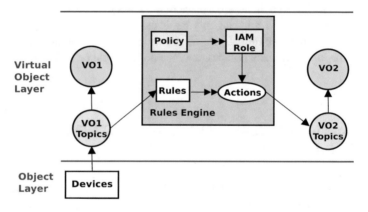

Fig. 5.10 The rules engine as a communication channel in AWS-IoT-ACMVO

The control point in AWS-IoT-ACMVO is located in rules engine to permit reserved topics of VO to communicate with other reserved topics of another VO through rules engine. For instance, once a data arrived to a reserved topic of a VO, a lambda function is triggered by a Lambda rule as an action only if the *Select* Clause and *Where* Clause in the SQL statement of the Lambda rule result is true. After the lambda function is triggered, an associated IAM role with Lambda function will react with an attached policy to authorize AWS-IoT to approach to the Lambda function and permit the Lambda function to perform actions with the arrived data. Lambda function can be allowed by IAM role policy to forward data to other reserved/unreserved topics. Hence, other topics can get data as long as the received data is in proper format without rejecting the received data or examining where data arrived from, and therefore, the received data will be forwarded to subscribers. Thus, a question such as "which resources should a topic receive data from?" is only checked through IAM role that is linked within an action of a rule. Topics have no choice over arrived data.

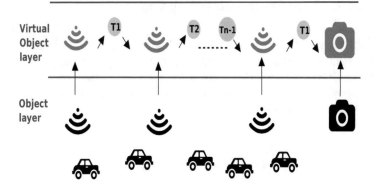

Fig. 5.11 The sensing speeding cars use case within ACO architecture [13]

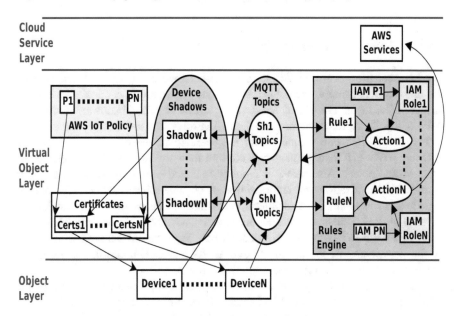

Fig. 5.12 The sensing speeding cars use case within AWS-IoT-ACMVO

A use case of sensing speeding cars is employed according to ACO-IoT-ACMsVO, within AWS-IoT-ACMVO. However, as mentioned earlier, access control points, communication style, and access control models are not exactly similar. Though AWS-IoT does not assist straight VO communication, the developed AWS-IoT-ACMVO allows indirect communication between VOs in the AWS-IoT. Therefore, the sensing speeding cars use case shown in Fig. 5.11 and employed within ACO architecture in [13], can be implemented and enforced within AWS-IoT as presented in Fig. 5.12. In the following section, the specifics of design, scenario, and authorization policy will be described.

5.6   A Use Case: Sensing Speeding Cars

In this part, two use cases of sensing cars' speed are presented. Both of the two scenarios contain sensors and a camera in the physical layer. Also, any object in the physical layer will push gathered data to the connected virtual objects (shadows). The two scenarios will show the communication among virtual objects and how this communication can be organized.

5.6.1   Sensing the Speed of a Single Car

The scenario and configuration of the first simple use case is as follows.

5.6.1.1   Setup and Configuration

The first scenario is a simplistic use case where there are only two physical sensors and one physical camera, each object is with one virtual object connected to it in the virtual object layer. The connected devices, virtual objects (shadows), certificates, AWS IoT policies, rules, actions and their IAM roles, and AWS services are shown in Fig. 5.13. First, one virtual object is generated for every physical object by utilizing AWS IoT management console and then link one X.509 certificate for every virtual object. Also, AWS IoT policy is linked to every certificate. In order to enable authentication and authorization of physical objects when they connect with their corresponding virtual objects, certificates are duplicated at their analogous physical objects. That is, specific actions (connect and publish) can be authorized by the linked AWS IoT policy for physical objects. Physical objects are accompanied by the private key of the certificate and an AWS root CA certificate, whenever certificates are granted to the corresponding physical objects.

Sensors and camera physical objects are simulated using AWS SDK for JavaScript (Node.js). For every MQTT update topic $aws/things/Sensor$_i$/sh adow/update there is an attached rule that triggers a Lambda function. As displayed in Fig. 5.13, Lambda functions are accountable for republishing the received stated data from the physical object (*Sensor$_i$* or *Camera*) to the following virtual object (*Virtual Sensor$_i$* or *Virtual Camera*). Moreover, every Lambda function is involved with an IAM role which permits AWS IoT to approach AWS and AWS IoT resources and services. The IAM role additionally can control Lambda function process such as obtaining the current state of a shadow or publishing data to other topics.

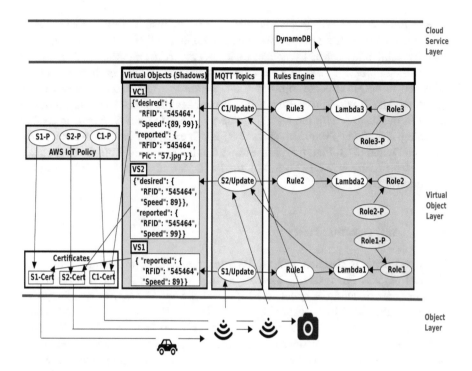

Fig. 5.13 A simple use case of sensing the speed of one car

5.6.1.2   Overall Scenario

Sensor₁ directs a *reported* message that contains RFID and Speed of the over speeding car to *Virtual Sensor₁* (*VS*1) by publishing to *Sensor₁* MQTT update topic $aws/things/Sensor₁/shadow/update. *Sensor₁* MQTT update topic will be attached to *Rule₁*, which will trigger *Lambda₁ function* each time data comes to *VS*1 MQTT update topic. *Lambda₁ function* republishes the received data to *Virtual Sensor₂* (*VS*2) with a *desired* tag. Figure 5.13 displays that the *reported* RFID and Speed to *VS*1 is republished to *VS*2 as *desired* by *Lambda₁ function*.

Sensor₂ sends RFID and Speed of the over speeding car in a *reported* message to *VS*2 by publishing to the MQTT topic $aws/things/Sensor₂/shadow/update. *Rule₂* is going to trigger *Lambda₂ function* every time data arrives to MQTT update topic of *VS*2. *Lambda₂ function* checks if the coming data is with *reported* tag, it compares the saved *desired* RFID with the incoming *reported* RFID from *Sensor₂*. If the two RFIDs match, *Lambda₂ function* combines the two speeds and one RFID and publishes it with *desired* tag to *Virtual Camera* (*VC*1). Figure 5.13 shows that the *reported* RFID matches the *desired* RFID in *Virtual Sensor₂*. Thus, we can see that *VC*1 has the combined speeds of the matched RFID with *desired* tag.

Virtual Camera will receive *reported* message from *Camera*, that contains RFIDs and pictures (Pic) of the passed cars. *Camera* will send the *reported* message by publishing to MQTT update topic $\mathtt{\$aws/things/Camera/shadow/}$ $\mathtt{update}$. $Lambda_3$ $function$ will be triggered by $Rule_3$ each time data is delivered to MQTT update topic of $VC1$. $Lambda_3$ $function$ examines if the incoming data is within a *reported* tag, it matches the stored incoming *desired* RFID from $Sensor_2$ with the incoming *reported* RFID from *Camera*. In case of matching, $Lambda_3$ $function$ adds the RFID, Speeds, and Pic and stores it in the Amazon DynamoDB. Figure 5.13 illustrates that the *reported* RFID in $VS1$ equals the *desired* RFID in $VC1$. Therefore, the RFID, Speeds, and Pic will be saved in the Amazon DynamoDB.

5.6.1.3 Authorization Policy

Every certificate is attached with an AWS IoT policy to permit certain actions for physical objects. For instance, $Sensor_1$ is only authorized to link and publish to $VS1$ in order to send the gathered RFIDs and Speeds of the over speeding cars. Therefore, the AWS IoT sensor2 policy ($S1$-P) and $VS1$ are linked with sensor1 certificate ($S1$-$Cert$), which is duplicated and copied to $Sensor_1$. The policy states that connect and publish actions are permitted to indicated resources, which in our scenario $VS1$ ($Sensor_1$ shadow). Likewise, $VS2$ and the AWS IoT sensor2 policy ($S2$-P) that is presented in Fig. 5.14 will be linked with $S2$-$Cert$ that is duplicated to $Sensor_2$, and the AWS IoT $C1$-P and $VC1$ will be linked with $C1$-$Cert$ that is duplicated to $Camera_1$.

AWS IoT describes policy variables that can be utilized in AWS IoT policies within the resource or condition block. The main variable $IoT : ClientID$ can be used to produce a single policy, which later can be linked to all certificates. However, certificates are not associated with an ID of physical sensor, which needs connect and publish to the attached shadows, so evil sensors might alter their ID to connect

Fig. 5.14 $S2$-P that is attached to $S2$-$Cert$

```
{ "Version": "2012-10-17",
  "Statement":
  [
    { "Effect": "Allow",
      "Action": [ "iot:Connect" ],
      "Resource": ["arn:aws:iot:us-west-2:760000000000:
      client/Sensor2"]
    },
    { "Effect": "Allow",
      "Action": [ "iot:Publish"],
      "Resource": [ "arn:aws:iot:us-west-2:76000000000:
      topic/$aws/things/Sensor2/shadow/update"]
    }
  ]
}
```

Fig. 5.15 $Role_2$ policy that
is attached to $Role_2$

```
{
    "Version": "2012-10-17",
    "Statement": [
        { "Effect": "Allow",
          "Action": "iot:GetThingShadow",
          "Resource": "arn:aws:iot:us-west-2:760000000000:
          thing/Sensor2"
        },
        { "Effect": "Allow",
          "Action": "iot:Publish",
          "Resource": "arn:aws:iot:us-west-2:760000000000:
          topic/$aws/things/Camera/shadow/update"
        }
    ]
}
```

and publish to another MQTT update topic. Thus, specifying a separate policy for each certificate is more secure as displayed in Fig. 5.13. Also, every AWS IoT policy is analogous to $S2$-P presented in Fig. 5.14 with regard to change sensor names.

Additionally, there is an IAM role coupled with every Lambda function to permit its access to AWS IoT resources and AWS services. For example, $Lambda_1$ is coupled with $Role_1$ to allow it publishing to the update topic of $VS2$.

Moreover, $Lambda_2$ is coupled with $Role_2$ to allow it receiving the *desired* state of $VS2$ and publishing to the $VC1$ MQTT update topic. The IAM $Role_2$ *Policy* that is coupled with $Role_2$ is shown in Fig. 5.15. Finally, $Lambda_3$ is authorized to receive the $VC1$ *desired* state and publish to Amazon DynamoDB by attaching $Role_3$ to $Lambda_3$.

5.6.2   Sensing the Speed of Multiple Cars

The first use case presents a simple idea of implementing the virtual object communication within AWS IoT and controlling the communication. Though, in real life, the need to track multiple cars appear, where many cars cross a sensor at same time. An AWS-IoT VO (shadow) has many reserved topics, which a VO subscribes to. When a sensor publishes a new list of RFIDs/Speeds, the previous list is removed and the new one is stored. However, historical data (old and new RFIDs) of the use case with multiple cars is required to be tracked and saved.

In this use case, for each VO analogous to a physical object, there is additional associated VO that works as historical storage. A way to push or receive data from the storage of VO is by employing a Lambda function that is triggered by publishing a sensor to an MQTT update topic of the analogous VO. Sensors ($S1, S2, \ldots,$ $Sn, C1$) and their analogous virtual objects ($VS1, VS2, \ldots, VSn, VC1$) and the storage for each of them ($VS1S, VS2S, \ldots, VSnS, VC1S$) is shown in Fig. 5.16.

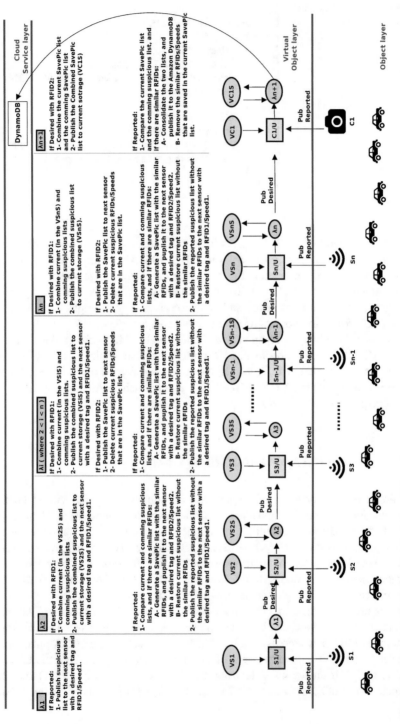

Fig. 5.16 A use case of sensing the speed of multiple cars

5.6.2.1   Setup and Configuration

As in the earlier use case, one virtual object and one virtual object storage are generated for each physical object, and then one X.509 certificate is linked for every virtual object. Certificates are duplicated and copied into their corresponding physical objects. Generally, certificates are attached with the AWS IoT policy, that states that sensors and the Camera are only permitted to connect and publish to the analogous VO.

Sensors and the Camera are simulated by employing AWS SDK for JavaScript (Node.js). Rules that are linked with MQTT update topics of VOs trigger Lambda functions. For example, rule1 that is linked to MQTT update topic of VS1 triggers $Lambda_1$. Generally, complex computations, such as obtaining the stored data, matching and consolidating the incoming and saved data, and republishing data to the current storage or next VO are the responsibility of Lambda functions. The functionality of every Lambda function is described in Fig. 5.16.

5.6.2.2   Overall Scenario

A *reported* message, which contains a list of RFIDs/Speeds of over speeding cars, is sent by $Sensor_1$ to *Virtual Sensor*$_1$ ($VS1$) by publishing to $VS1$ MQTT update topic. $Lambda_1$ $function$ is triggered by $Rule_1$ whenever a published request reaches to MQTT update topic. Figure 5.16 describes how $Lambda_1$ $function$ will deal with the reported data.

A *reported* message, which contains a list of RFIDs/Speeds of over speeding cars, is sent by $Sensor_2$, ..., $Sensor_i$, .., $Sensor_n$ to their analogous VO by publishing to VO_i MQTT update topic, where $2 \leq i \leq n$. $Lambda_i$ $function$ is triggered by $Rule_i$ whenever a published request reaches to MQTT update topic of VS_i. Figure 5.16 describes how $Lambda_i$ $function$ will deal with the received data. Note that $Lambda_3$ to $Lambda_{(n-1)}$ will perform alike computations.

5.6.2.3   Authorization Policy

As in the earlier use case, an AWS IoT $Policy$ is linked with $S1 - Certificate$, .., $Sn - Certificate$, $C1 - Certificate$. The policy states that physical objects are authorized to connect, and publish are authorized to their analogous VO, and physical objects are authorized to publish to MQTT update topic of their analogous VO. An example of an associated policy that permits only connecting and publishing is presented in Fig. 5.14.

Moreover, the IAM roles are associated with Lambda functions. For instance, $Role_1$ is attached to $Lambda_1$ to permit it publishing to the update topic of $VS2$. $Role_2$ is attached to $Lambda_2$ to permit it receiving data of the $VS2$ storage, and to authorize it publishing to the update topics of $VS2S$ and $VS3$. $Role_{(n+1)}$ is attached to $Lambda_{(n+1)}$ to permit it receiving data of the $VC1$ storage, and to authorize it

Fig. 5.17 $Role_5$ that is
attached to $Lambda_5$

```
{"Version": "2012-10-17",
   "Statement": [
     {  "Effect": "Allow",
        "Action": "iot:Publish",
        "Resource": "arn:aws:iot:us-west-2:760000000000:
         topic/$aws/things/Sensor5_Storage/shadow/update"
     },
     {  "Effect": "Allow",
        "Action": "iot:GetThingShadow",
        "Resource": "arn:aws:iot:us-west-2:769000000000:
         thing/Sensor5_Storage"
     },
     {  "Effect": "Allow",
        "Action": "iot:Publish",
        "Resource": "arn:aws:iot:us-west-2:769000000000:
         topic/$aws/things/Camera/shadow/update"}
   ]
}
```

publishing only to its own storage and then to the Amazon DynamoDB. $Role_5$ that is attached to $Lambda_5$ is shown in Fig. 5.17. $Role_5$ permits $Lambda_5$ to receive data which is stored in the $VS5$ storage (the implementation assume $n = 5$) and publish only to the update topics of $VS5S$ and $VC1$.

5.6.3   Performance Evaluation and Discussion

Our scenarios disseminate the Suspicious list published by any sensor till the ultimate virtual sensor, and similarly disseminate the *SavePic* list from the point of creation till the camera. The initial likely created Suspicious list begins from $Sensor_1$, and the initial likely created *SavePic* list begins when $Sensor_2$ publishes comparable Suspicious list to the published Suspicious list by $Sensor_1$. In this part, the time of propagating the *SavePic* list and the Suspicious to their ending destination is computed.

The multiple sensors and cars use case is used in calculating the propagation time. The number of sensors is set to five. Two AWS SDKs is used for JavaScript (Node.js) to subscribe to $Virtual\ Sensor_5\ Storage\ (VS5S)$ and $Virtual\ Camera_1\ Storage\ (VC1S)$, thus an acknowledgement can be received when the Suspicious and the *SavePic* list arrive. A bash script is created to execute $Sensor_1$, start the timer, run $VS5S$, and end the timer when an acknowledgement is received from $VS5S$. Similarly, the bash script will execute $Sensor_2$ (with alike RFIDs of $Sensor_1$), start the timer, run $VC1S$, and end the timer when an acknowledgement is received from $VC1S$. Thus, the propagation time of the Suspicious and the *SavePic* list to their final destination is evaluated.

In order to publish the Suspicious list with 1, 10, 20, 30, and 40 RFIDs, $Sensor_1$ is run. For the Suspicious list that has one RFID, the average propagation time of 10 runs is calculated. In Fig. 5.18, the average propagation time of the Suspicious list with one RFID from S_1 till $VS5S$ is 5915 milliseconds, which is the average of 10

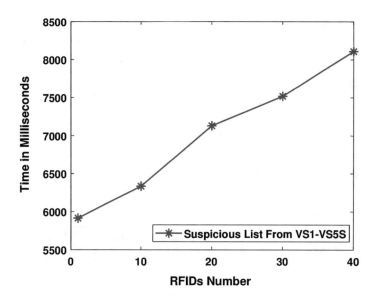

Fig. 5.18 Propagation time of suspicious list from S_1 until $VS5S$

runs. Also, the average propagation time of the Suspicious list with 10, 20, 30 and 40 RFIDs from S_1 till $VS5$ is respectively 6335, 7131, 7519, and 8109 milliseconds, which are the average of 10 runs. Note that the outliers are eliminated, such as time values that exceed 10000 or less than 3000 milliseconds.

When a Suspicious list is published by S_1 and an acknowledgement is arrived from $VS5S$, then $Sensor_2$ is also executed to publish a Suspicious list that is similar to Published Suspicious list by S_1 with 1, 10, 20, 30 and 40 RFIDs. Figure 5.19 shows the average propagation time of the SavePic list with 1, 10, 20, 30 and 40 RFIDs from S_2 until $VC1S$ is respectively 7774, 8100, 8405, 8694 and 8851 milliseconds, which are the average of 10 times run. Note that the outliers are eliminated, which is values that less than 4000 or exceed 14,000 milliseconds.

The use case in Fig. 5.16 has extra computation and steps when a Lambda function receives the Suspicious list than when a Lambda function receives the SavePic list. However, in Figs. 5.18 and 5.19 results, the propagation time of the SavePic lists are more than the propagation time of the Suspicious lists. Different values due to the bigger payload of the SavePic list, which has two speeds for every RFID, than the Suspicious list, which has only single speed for every RFID.

5.7 Summary

This chapter presented a discussion of developing operational access control models for VO communication, administrative access control models for VO communica-

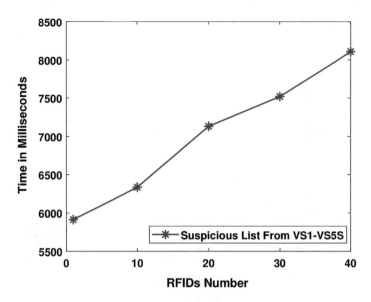

Fig. 5.19 Propagation time of SavePic list from S_2 until $VC1S$

tion, and AWS-IoT-ACMVO model for AWS-IoT VO (shadows) communication. These access control models are developed upon the access-control oriented (ACO) for cloud-enabled IoT architecture presented in Chap. 2, with VOs and cloud services in the middle layers. A central aspect of ACO is to control communication among VOs. Thus, these models in this chapter will act as guidelines for researchers to recognize access control decision and enforcement points essential in VOs communication. It helps applying control points on virtual object communication to secure accessing to VO and topics. Real-world use cases of sensing cars's speed with single and multiple cars scenarios and access control requirements reflect the need and use of access control framework for VO communication in AWS-IoT. The following chapter will discuss the fine grained access and communication control models proposed and implemented in different CPS and IoT domains.

References

1. Ouaddah, A., Mousannif, H., Elkalam, A. A., & Ouahman, A. A. (2017). Access control in the internet of things: Big challenges and new opportunities. *Computer Networks, 112*, 237–262.
2. Gusmeroli, S., Piccione, S., & Rotondi, D. (2013). A capability-based security approach to manage access control in the internet of things. *Mathematical and Computer Modelling, 58*(5), 1189–1205.
3. Mahalle, P. N., et al. (2013). Identity authentication and capability based access control (iacac) for the internet of things. *Journal of Cyber Security and Mobility, 1*(4), 309–348.

4. Alshehri, A., & Sandhu, R. (2016a). Access control models for cloud-enabled internet of things: A proposed architecture and research agenda. In *2016 IEEE 2nd International Conference on Collaboration and Internet Computing (CIC)* (pp. 530–538). IEEE.

5. Nitti, M., Pilloni, V., Colistra, G., & Atzori, L. (2015). The virtual object as a major element of the internet of things: A survey. *IEEE Communications Surveys & Tutorials, 18*(2), 1228–1240.

6. How the AWS IoT Platform Works—AWS. https://aws.amazon.com/iot/how-it-works/. [Online; accessed 04-March-2021].

7. Bhatt, S., Patwa, F., & Sandhu, R. (2017). Access control model for aws internet of things. In *International Conference on Network and System Security* (pp. 721–736). Springer.

8. Zhang, Y., Patwa, F., & Sandhu, R. (2015). Community-based secure information and resource sharing in aws public cloud. In *2015 IEEE International Conference on Collaboration and Internet Computing (CIC)* (pp. 46–53). IEEE.

9. Sandhu, R. S. (1998). Role-based access control. In *Advances in Computers* (vol. 46, pp. 237–286). Elsevier.

10. Jin, X., Krishnan, R., & Sandhu, R. (2012). A unified attribute-based access control model covering dac, mac and rbac. In *IFIP Annual Conference on Data and Applications Security and Privacy* (pp. 41–55). Springer.

11. Gupta, M., & Sandhu, R. (2016). The GURA$_G$ administrative model for user and group attribute assignment. In *International Conference on Network and System Security* (pp. 318–332). Springer International Publishing.

12. Alshehri, A., & Sandhu, R. (2016b). On the relationship between finite domain ABAM and PreUCON$_A$. In *International Conference on Network and System Security* (pp. 333–346). Springer.

13. Alshehri, A., & Sandhu, R. (2017). Access control models for virtual object communication in cloud-enabled IoT. In *2017 IEEE International Conference on Information Reuse and Integration (IRI)* (pp. 16–25). IEEE.

14. Alshehri, A., Benson, J., Patwa, F., & Sandhu, R. (2018). Access control model for virtual objects (shadows) communication for aws internet of things. In *Proceedings of the Eighth ACM Conference on Data and Application Security and Privacy* (pp. 175–185).

15. Ferraiolo, D. F., Sandhu, R., Gavrila, S., Kuhn, D. R., & Chandramouli, R. (2001). Proposed nist standard for role-based access control. *ACM Transactions on Information and System Security (TISSEC), 4*(3), 224–274.

16. Sandhu, R. S., Coyne, E. J., Feinstein, H. L., & Youman, C. E. (1996). Role-based access control models yz. *IEEE Computer, 29*(2), 38–47.

17. Sandhu, R., Bhamidipati, V., & Munawer, Q. (1999). The arbac97 model for role-based administration of roles. *ACM Transactions on Information and System Security (TISSEC), 2*(1), 105–135.

18. Bacon, J., Eyers, D. M., Singh, J., & Pietzuch, P. R. (2008). Access control in publish/subscribe systems. In *Proceedings of the Second International Conference on Distributed Event-Based Systems* (pp. 23–34).

19. Gupta, M., Patwa, F., & Sandhu, R. (2018). An attribute-based access control model for secure big data processing in hadoop ecosystem. In *Proceedings of the Third ACM Workshop on Attribute-Based Access Control* (pp. 13–24).

20. Gupta, M., Benson, J., Patwa, F., & Sandhu, R. (2020). Secure V2V and V2I communication in intelligent transportation using cloudlets. *IEEE Transactions on Services Computing*. https://doi.org/10.1109/TSC.2020.3025993.

21. Gupta, M., Benson, J., Patwa, F., & Sandhu, R. (2019). Dynamic groups and attribute-based access control for next-generation smart cars. In *Proceedings of the Ninth ACM Conference on Data and Application Security and Privacy* (pp. 61–72).

22. Cathey, G., Benson, J., Gupta, M., & Sandhu, R. (2021). Edge centric secure data sharing with digital twins in smart ecosystems. *Preprint arXiv:2110.04691*.

23. Gupta, M., & Sandhu, R. (2021). Towards activity-centric access control for smart collaborative ecosystems. In *Proceedings of the 26th ACM Symposium on Access Control Models and Technologies* (pp. 155–164).

24. Bhatt, S., Pham, T. K., Gupta, M., Benson, J., Park, J., & Sandhu, R. (2021). Attribute-based access control for AWS internet of things and secure Industries of the Future. *IEEE Access, 9*, 107200–107223.
25. Gupta, D., Bhatt, S., Gupta, M., Kayode, O., & Tosun, A. S. (2020). *Access control model for google cloud IoT. 2020 IEEE 6th Intl conference on big data security on cloud (BigDataSecurity), IEEE Intl conference on high performance and smart computing, (HPSC) and IEEE Intl conference on intelligent data and security (IDS)* (pp. 198–208). IEEE.
26. Gupta, M., Awaysheh, F. M., Benson, J., Alazab, M., Patwa, F., & Sandhu, R. (2020). An attribute-based access control for cloud enabled industrial smart vehicles. *IEEE Transactions on Industrial Informatics, 17*(6), 4288–4297.
27. Gupta, M., Patwa, F., & Sandhu, R. (2017). POSTER: Access control model for the Hadoop ecosystem. In *Proceedings of the 22nd ACM on Symposium on Access Control Models and Technologies* (pp. 125–127).
28. Gupta, M., Patwa, F., & Sandhu, R. (2017). Object-tagged RBAC model for the Hadoop ecosystem. In *IFIP Annual Conference on Data and Applications Security and Privacy* (pp. 63–81). Springer.

Chapter 6
Attribute Based Access Control for Intelligent Transportation

6.1 Introduction

Future connected world incorporate IoT and CPS technologies where domains such as smart connected cars and intelligent transportation will be the key drivers. These critical ecosystems will involve different smart entities including the traffic lights, gas stations, emergency vehicles, ambulances, or any connected vehicle which will connect and communicate with each other offering plethora of applications and services to the end users. The idea of connected vehicles talking to each other (vehicle to vehicle—V2V) or traffic infrastructure interacting and exchanging messages with the vehicles (vehicle to infrastructure—V2I) will evolve using cloud and edge enabled technologies. These smart vehicles will receive different notification, alerts and services using technologies such as Dedicated Short Range Communication (DSRC) protocol in Basic safety messages (BSMs), which will display these alerts on the dashboard, or may even vibrate the car seat. Vehicles will receive information from restaurants, gas stations, or parking garages in their location vicinity which will support pleasant travel experience as well make the traffic flow more efficient and environment friendly. These smart cars have more than 100 million lines of code and several hundreds of electronic control units (ECUs) and sensors which generate humongous amount of data, referring them often as *datacenter on wheels*.

These usually isolated physical entities are now exposed to the cyber world making them vulnerable to cyber attacks. Several common vulnerabilities such as malware, over-privilege, buffer overflow etc. have been demonstrated in the past by researchers. However, with respect to the intelligent transportation concerns like fake and untrusted messages, sensors and ECU remote control, spoofed messages from an adversary vehicle, or a even a faulty sensor can be common issues which must be addressed. In addition, the broader attack surface due to different applications and smart sensors can expose critical units of a vehicle against malicious actors. Further, interaction with different entities externally such as gas station, toll booths, parking garages, etc. are a big attack vector for the connected smart car security. Researchers have orchestrated cyber attacks on smart vehicles including Tesla model X[1] and Jeep,[2] which resulted in stopping engine and remotely controlling the car steering. These incidents will grow as more connected vehicles are rolled-out on the road which can have implications beyond monetary concerns and are threat to life and limb.

Authorization and access control [1–10] security solutions are used to limit the rogue and malicious access and control to resources in an enterprise. Similar access control mechanisms are also needed in IoT and CPS like ecosystem (as discussed in earlier chapters) to limit control and interaction among different smart entities. Attribute based access control offers fine grained and flexible solution in multi-domain distributed scenarios where objects are physically scattered and controlled by different administrators. Intelligent transportation requires authorized dynamic communication and data exchange within different smart objects. It is also important to ensure that only authorized entities must be allowed to access and control on-board ECUs, sensors, read data or send notifications and alerts within the ecosystem. The solution proposed in this chapter considers attributes of different moving smart objects such as location, vehicle type, speed, directions etc. to dynamically allocate them to different groups which are created in limited geographic area by the administrators. In addition, user specified privacy preferences and personal requirements will also be defined so as to ensure the relevance of different notifications, alerts and advertisements which can be received based on changing and fluid mobile smart vehicles. The model is implemented using Amazon Web Services (AWS) IoT platform and has the flexibility to be deployed in any cloud service provider platforms.

6.1.1 Chapter Organization

This chapter will provide the foundational access control model for smart cars. In Sect. 6.2, we will discuss the authorization requirements for intelligent transportation and connected smart cars, emphasizing the need for multi-layer security with user privacy policies along with the relevance of groups in locations centric smart transportation ecosystem. In Sect. 6.3, we discuss our formal dynamic groups and ABAC model referred as CV-ABAC$_G$. In Sect. 6.4, we provide AWS enforcement of our proposed model discussing the use cases along with prototype implementation and performance metrics. Section 6.5 concludes this chapter.

---

[1]https://www.usatoday.com/story/tech/2017/07/28/chinese-group-hacks-tesla-second-year-row/518430001/.

[2]https://www.wired.com/2015/07/hackers-remotely-kill-jeep-highway/.

6.2 Authorization Requirements in ITS

Connected vehicles expose the usually isolated vehicles to an external environment using different protocols such as LTE, 5G, WiFi, LoRA, DSRC etc. V2V and V2I communication results in very short lived and dynamic interaction among different entities in surroundings, which require confidentiality and integrity of messages. In this section, we will discuss the authorization needs along with relevance of dynamic groups in Intelligent Transportation System (ITS).

6.2.1 Multi-Layer and User Privacy Preferences

The externally exposed attack surface of smart cars offers the preliminary entry point to compromise the in-vehicle systems. It is important to have multi-level access control security policies to secure external interfaces and in-vehicles CAN based ECU communication. Authorization for external ecosystem will offer protection for on-board sensors, data and applications from unauthorized communication and control by various entities such as traffic lights, cloud-based applications, or masquerading adversaries posing as mechanic. Over the air updates must be checked and are only allowed from legit sources or vehicle manufacturer. An adversary who is successful in by passing the initial security check can be restricted by in-vehicle security level which is deployed to secure control and overwrite of critical vehicle units such as brakes, transmission or engine systems. Vehicles can exchange basic safety messages [11–14] which must be trusted. Before the exchanged information is used by applications in the vehicles, it is needed to ensure that the data and information received is true, confidential and from a trusted source. Different applications access devices and sensors inside and outside the vehicle, such as a lane departure warning system which access sensors on the tire, must be secure and authorized to restrict malicious applications reading and controlling vehicle movement. A passenger accessing infotainment (information and entertainment) systems of the car via Bluetooth or using smartphone inside car must also be authorized.

In ITS location-based services support alerts and notification for various promotions, flash flood or other scenarios. Users must have the capability to define their privacy preferences to decide what kind of advertisements they want to receive, and filter the ones which are not satisfying the user needs. As an example, a user would like to have notifications from different restaurants in its current location vicinity, but is not interested to get optimal routes messages due to traffic jams. As the same time, there are system wide policies which are applicable to the entire ITS such as what is the speed limit, or who can issue tickets on a speeding smart vehicle.

6.2.2   Relevance of Groups

Many ITS applications and value-added services needed from drivers are very location centric and time sensitive. As an example, a driver would want to receive warnings near school zone, crossing pedestrians or in case of a blind spot. In addition, the alerts and notifications are dependent on the GPS locations of the vehicle and are usually short-lived. A gas station may send discount notifications to nearby vehicles, ice on the bridge alerts to vehicles heading in the direction of the bridge, or deer threat in the area, are some examples where notifications and warnings can be send to all the vehicles in close vicinity. Hence, dynamic segregation of different connected vehicles in to various location and other types of groups can help in limiting which vehicles will be notified and also helps in easy administration as one notification for the entire group can send alerts to all its vehicle members at the time of alert. In addition, smart vehicles which are present at a location can have similar characteristic such as deer-threat alert, speed limit, flash-flood, etc. which they can inherit from the group to which they are dynamically assigned based on their movement and other factors. As represented in Fig. 6.1, different smart entities are categorized into various location groups which are pre-defined and demarcated by city authorities in a smart city environment. Such groups can be dynamically allocated to smart vehicles depending on the attributes, personal needs, requirements and interest or current location.

In addition, group hierarchy can also be created as illustrated in Fig. 6.2 where child groups can be created in larger parent groups which will fine grain the notifications to only a subset of vehicles. As an example, within a location group different subgroup can be developed based on vehicle type such as bus, car, ambulance, trailers etc. to target alerts only to specific category of vehicles within a larger location group. Further, groups can also be created for different service type, such as a group for vehicle which want to part of car-pooling service, or a group for smart cars which want to receive restaurant notifications. This group hierarchy support attributes inheritance where child groups receive attributes from parent.

6.3   Dynamic Groups and ABAC Model

Data exchange and interaction among various smart entities in a connected ITS need policies at various levels which can be administered centrally by a single administrator, along with individual user privacy preferences. Hence, an access control model should cover all the requirements of users and system wide at the same time offer fine grained security mechanisms. In this section, we will define and formulate a mathematically grounded attribute based access control model supporting dynamic groups for connected vehicles, referred to as CV-ABAC$_G$.

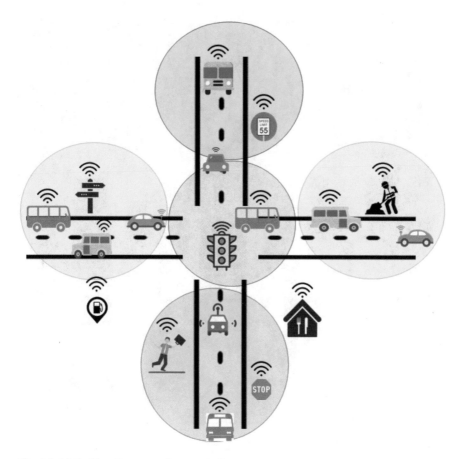

Fig. 6.1 Marked location groups in a smart city

6.3.1   CV-ABAC$_G$ Model Overview

Figure 6.3 shows the conceptual CV-ABAC$_G$ model followed by mathematical formal notations in Table 6.1. The basic components of the model include Sources (S), Clustered Objects (CO), Objects in clustered objects (O), Groups (G), Operations (OP), Activities (A), Authorization Policies (POL), and Attributes (ATT).

Sources (S) Activities are initiated by these entities on other smart objects, dynamic groups and various other applications in the system. Some examples of a source include a user, sensor, application, service, administrator, clustered object like smart vehicles, or various groups created in the system. As an example, a deer threat or flash flood warning can be triggered by the police car, or road side sensor which act as a source of the activity. On the other hand, when a mechanic or

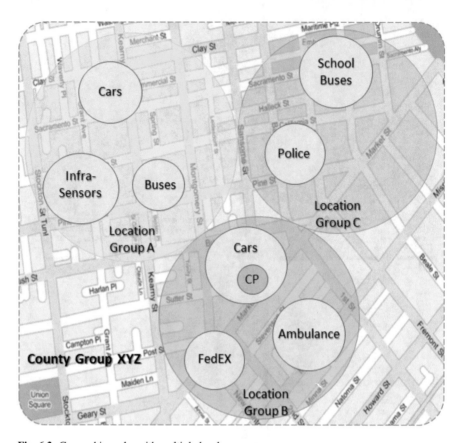

Fig. 6.2 Groups hierarchy with multiple levels

manufacturer issues a command on the vehicle, or a gas station sends coupons and advertisements alerts they are considered a source in the system.

Clustered Objects (CO) In ITS these clustered objects are extremely relevant, which have several sensors and actuators within them similar to a connected vehicle, traffic light or other smart objects. For example, a smart car has multiple on-board sensors including lane departure, blind spot, tire pressure, telematics, transmission and engine control ECUs along with multitude of applications. These connected COs communicate with each other and exchange information among them or with remote sources such as cloud based applications and services. These clustered objects also reflect V2V and V2I communication using the basic safety messages.

Objects in Clustered Objects (O) The objects include ECUs, sensors and different applications which are installed in a clustered object. Such objects cover the internal state of the vehicles such as the emission control, engine diagnostics, cabin monitoring system as well as sensors from external environment including temperature, rain, tire, cameras etc. Commands can be issued on these objects to

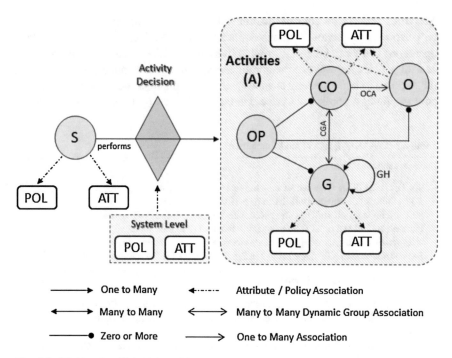

Fig. 6.3 CV-ABAC$_G$ conceptual model

read data, as well as to trigger an action on them. Applications such as lane departure warning or tire pressure can also have access to various sensors in the vehicle so as to provide alerts to users or providing diagnostic data to remote services in cloud.

Group (G)   Logical collection of different clustered object having similar features and needs is called a group. These groups can support focused notifications to a subset of vehicles and also enable assignment of attributes to member of the groups. These groups can be created based on the location, services preferences such as gas station alerts, car-pooling notifications etc. or type of vehicles such as bus, truck, car etc. Group hierarchy (GH) can be created which supports inheritance of attributes and security policies from parent group to child groups. In our model, we limit a smart vehicle or CO to be a direct member of only one group which are at the same level. This helps in managing the attributes inheritance and support the practical feasibility of our model.

Operations (OP)   Groups, clustered objects or other smart entities can perform different actions referred as operations. This include from simple read or writing sensors data to the ECU, triggering alerts in the system or administrative operations as well such as adding new policies or attributes that can be performed by administrators.

Activities (A) These consider administrative and operational activities that can be performed by various sources defined in the ITS. Single or multiple atomic operations (OP) can be a part of an activity that will also require security access control policies for user personal preferences, system wide policies or can be both, for deciding to allow or deny an activity. As an example, a source requesting a car-pool generates a pooling alert that can be broadcasted to different vehicles which

Table 6.1 CV-ABAC$_G$ formal model definitions

Basic sets and functions

- S, CO, O, G, OP are finite sets of sources, clustered objects, objects, groups and operations.
- A is a finite set of activities which can be performed in system.
- ATT is a finite set of attributes associated with S, CO, O, G and system-wide.
- For each attribute att in ATT, Range(att) is a finite set of atomic values.
- attType: ATT = {set, atomic}, defines attributes to be set or atomic valued.
- Each attribute att in ATT maps entities in S, CO, O, G to attribute values. Formally,

$$\text{att} : S \cup CO \cup O \cup G \cup \{\text{system-wide}\} \rightarrow \begin{cases} \text{Range(att)} \cup \{\perp\} & \text{if attType(att) = atomic} \\ 2^{\text{Range(att)}} & \text{if attType(att) = set} \end{cases}$$

- POL is a finite set of authorization policies associated with individual S, CO, O, G.
- directG : CO → G, mapping each clustered object to a system group, equivalently CGA ⊆ CO × G.
- parentCO : O → CO, mapping each object to a clustered object, equivalently OCA ⊆ O × CO.
- rootparentG : G → G, mapping each group to root parent group in hierarchy.
- GH ⊆ G × G, a partial order relation $\succeq_g$ on G. Equivalently, parentG : G → 2^G, mapping group to a set of parent groups in hierarchy.

Effective attributes of groups, clustered objects and objects (derived functions)

- For each attribute att in ATT such that attType(att) = set:
  - effG$_{att}$: G → $2^{\text{Range(att)}}$, defined as effG$_{att}$(g$_i$) = att(g$_i$) ∪ ($\bigcup_{g \in \{g_j | g_i \succeq_g g_j\}}$ effG$_{att}$(g)).
  - effCO$_{att}$: CO → $2^{\text{Range(att)}}$, defined as effCO$_{att}$(co) = att(co) ∪ effG$_{att}$(directG(co)).
  - effO$_{att}$: O → $2^{\text{Range(att)}}$, defined as effO$_{att}$(o) = att(o) ∪ effCO$_{att}$(parentCO(o)).

- For each attribute att in ATT such that attType(att) = atomic:
  - effG$_{att}$: G → Range(att) ∪ {⊥},

    $$\text{defined as effG}_{att}(g_i) = \begin{cases} \text{att}(g_i) & \text{if } \forall g' \in \text{parentG}(g_i). \text{ effG}_{att}(g') = \perp \\ \text{effG}_{att}(g') & \text{if } \exists \text{ parentG}(g_i). \text{ effG}_{att}(\text{parentG}(g_i)) \neq \perp \text{ then select} \\ & \text{parent } g' \text{ with effG}_{att}(g') \neq \perp \text{ updated most recently.} \end{cases}$$

  - effCO$_{att}$: CO → Range(att) ∪ {⊥},

    $$\text{defined as effCO}_{att}(co) = \begin{cases} \text{att}(co) & \text{if effG}_{att}(\text{directG}(co)) = \perp \\ \text{effG}_{att}(\text{directG}(co)) & \text{otherwise} \end{cases}$$

  - effO$_{att}$: O → Range(att) ∪ {⊥},

    $$\text{defined as effO}_{att}(o) = \begin{cases} \text{att}(o) & \text{if effCO}_{att}(\text{parentCO}(o)) = \perp \\ \text{effCO}_{att}(\text{parentCO}(o)) & \text{otherwise} \end{cases}$$

(continued)

Table 6.1   (continued)

Authorization functions (policies)

- Authorization Function: For each op $\in$ OP, Auth_{op}(s : S, ob : CO $\cup$ O $\cup$ G) is a propositional logic formula returning true or false, which is defined using the following policy language:

  - $\alpha ::= \alpha \wedge \alpha \mid \alpha \vee \alpha \mid (\alpha) \mid \neg\alpha \mid \exists\, x \in \text{set}.\alpha \mid \forall\, x \in \text{set}.\alpha \mid \text{set } \triangle \text{ set} \mid \text{atomic} \in$ set $\mid$ atomic $\notin$ set
  - $\triangle ::= \subset \mid \subseteq \mid \nsubseteq \mid \cap \mid \cup$
  - set $::= \text{eff}_{att}(i) \mid att(i)$ for att $\in$ ATT, i $\in$ S $\cup$ CO $\cup$ O $\cup$ G $\cup$ {system-wide}, attType(att) = set
  - atomic $::= \text{eff}_{att}(i) \mid att(i) \mid$ value   for att $\in$ ATT, i $\in$ S $\cup$ CO $\cup$ O $\cup$ G $\cup$ {system-wide}, attType(att) = atomic

Authorization decision

- A source s $\in$ S is allowed to perform an activity a $\in$ A, stated as Authorization(a:A, s:S), if the required policies needed to allow the activity are included and evaluated to make final decision. These multi-layer policies must be evaluated for individual operations (op_i $\in$ OP) to be performed by source s $\in$ S on relevant objects ($x_i \in$ CO $\cup$ O $\cup$ G). Formally, Authorization(a:A, s:S) $\Rightarrow \text{Auth}_{op_1}$ (s : S, x_1), Auth_{op_2} (s : S, x_2), ..., Auth_{op_n} (s : S, x_3)

are in the geographic area, however, only some drivers may actually be able to see the request whose individual policies are satisfied. As such, a user may not want to get the car-pooling notification from requestor who got very low ratings. Hence, multiple policies can be involved in an activity defined at different levels that must be enforced to cover both system wide and personal preferences. These activities have been categorized into following.

Service Requests These activities are initiated by the users and other entities. As an example, a service request is generated when a vehicle breaks-down or a car-pool request by a user to all the vehicles which satisfy personal policies.

Administration Administrative operations can be performed which include modifying security policy, changing the group hierarchy or attributes of various entities. The scope of the group can be defined, how smart vehicles can be assigned to various groups, using the personal preferences etc.

Notifications Various members in the groups can be notified for different updates regarding the group such as flash flood warning, deer threat alerts, or for getting promotions based on locations for garage parking or restaurants.

Control and Usage Read, control and write are some of the operations which can be performed in the vehicle or remotely. A car manufacturer issuing patches or mechanic reading engine data are remote activities. A passenger in the car using his smart phone to use infotainment and on-board vehicle applications are local control activities.

Security Policies and Attributes Our proposed CV-ABAC$_G$ model covers privacy policies from different objects and entities and allow the capability to specify such policies using their attributes. As in Fig. 6.3, source can have its own set of personal policies along with different attributes it can have such as gender, age, location, speed, direction etc. Clustered objects can have its own policy defined, for example, a vehicle can define which mechanics can access its data, or what notification it may want to receive etc. These COs also can have their own attributes such as vehicle size, type, or GPS locations. Various groups created in the ecosystem can have set of attributes and policies, for instance, restaurant notification group can define who can join it. System wide policies can be created to determine how groups can send notifications and alerts to various members who are part of the activity. A clustered object can be dynamically assigned to various groups as they move along various locations or changing requirements, that will support the inheritance of attributes from groups to its members. These attributes of various entities can change more frequently as compared to policies defined in the system. These attributes can be added or removed more dynamically as the vehicles move or change their surroundings such as accident, location, traffic situations etc. These policies are defined by the administrators which are more fixed aka static in nature and only the attributes which are part of the policy will define its outcome. As an example, a driver stating a policy that "Only Panda express is allowed to send notification between 6-9 pm". In this requirement, the attributes of the source restaurant are checked and if that matches with the desired restaurant name in the policy defined by the user, then it will be pushed to the car dashboard or so. In addition, dynamic policies can also be defined which allows to specify, for example, ambulance in particular locations are alerted in case of medical emergency. However, this can change when a bigger health care issue is noticed, and such policy can be dynamically overwritten based on the gravity of the situation. It must be noted that policy can not be changed at the time of activity evaluation. Any change to the activity may be reflected in subsequent policy evaluation. Multi-level policies can also be evaluated including the user privacy policies.

6.3.2 Components Definitions

Table 6.1 shows how the different entities such as sources, objects, clustered objects can be assigned different values for the attributes from the Range(att) of each of these attributes att from the set ATT. These attributes can be either set or atomic valued based on the attType function and the kind of attribute it is. Different entities can also be assigned null for atomic values attribute or may also have multiple values for different set valued attributes from its range. The set of security policies associated with each entities is reflected by the POL.

Different groups defined in the ecosystem can have multiple clustered groups as its members depending on their needs and personal preferences. As an instance, a vehicle can be associated with a location based group depending on its GPS. Our model supports the fact that a vehicle or clustered object can only be associated with one group at the same level of hierarchy as defined by the direct function. As child groups inherit attributes from various parent groups, it is enough to assign a clustered object to one the groups at the same level and still support inheritance of appropriate attributes. As these clustered objects have different smart objects, sensors and applications installed, these can also be accessed by various other sources in the system. The parentCO function assigns an object to its corresponding clustered object in a one to many mapping, stating that one object can only be associated with one clustered object whereas a clustered object may have many objects inside it. In addition, GH illustrated by the self-loop, reflects group hierarchy formally denoted by a partial ordered relation on set G. This partial order relation is stated as $\succeq_g$ such that $g1 \succeq_g g2$ meaning g1 is a child group which inherits all the attributes from the parent group g2 which is computed using the parentG function.

These groups bring in the benefit of easy administration since several attributes are assigned or can be removed from its members just with one administrative operation. In addition, groups hierarchy supports inheritance of attributes, hence for the attributes which are set valued type, effective attributes of att for a group g1 as denoted using $effG_{att}(g1)$ is the union of values which are directly assigned to attribute att and the effective values which are inherited from all the parent groups as defined in the hierarchy. As such for clustered objects, the effective attributes for att as defined by $effCO_{att}$ can be the attributes which are directly associated for att along with the attribute values which are effective for att from the groups to which the CO is a member calculated with directG. Apart from the direct attributes, objects in the clustered objects will also be inherit the attributes from the clustered object which can be calculated with $effO_{att}$. The union operation is used in case of set valued attributes which will not work for atomic valued since it can only hold one value. There are several ways to address this problem in atomic valued. In our approach, for groups, we update the recent non-null attribute value from the parent groups which will overwrite the attribute in child groups as shown in Table 6.1. Not-null atomic values from groups can be inherited for clustered objects as define by $effCO_{att}(co) = effG_{att}(directG(co))$. For objects, the non null atomic attribute for the clustered object will overwrite the objects value. For atomic attributes, if the parent(s) has null value for an attribute, the entity (group, clustered object or object) will retain its directly assigned value without any overwrite.

Each operation will have an authorization function that specify the policies which need to be satisfied to allow the operation op $\in$ OP. $Auth_{op}(s : S, ob : CO \cup O \cup G)$ is the authorization function in the set POL that defined the conditions to allow a source s to perform an operation op on objects which can be objects, clustered objects or groups in the system. These security policies contain the user personal policies defined for clustered objects, objects and various groups that also include system wide policies defined by the security administrators. These condition in the policies can be defined using propositional logic based language as stated in Table 6.1. A set of policies must be complied so as to allow an activity in the system. An authorization function for activity a $\in$ A requested from a source s defined as

Authorization(a : A, s : S), contains different system wide as well as user privacy policies to be allowed to have an activity a from source s.

The CV-ABAC$_G$ model offers fine grained policy specification and enforcement which fits into the dynamic and mobile intelligent transportation. The model supports location sensitive and time focused services and applications supporting personal privacy preferences along with system wide policies. The model introduces the novel notion of dynamic groups to support easy administration of the notification and alerts in the system.

6.4   AWS Enforcement

This section presents a prototype implementation of our proposed CV-ABAC$_G$ model demonstrating the intelligent transportation use case deployed in AWS IoT platform. This proof of concept will reflect the dynamic assignment of vehicle in groups as a part of administrative aspect along with the multi level security policy need supported by the AWS. The experiments were simulated to showcase real smart vehicles without compromising the applicability and performance evaluation of the deployed model. There was no long term data collection in the central cloud to limit any privacy concerns when sharing location and other user data with the cloud.

6.4.1   Use Case Overview

In intelligent transportation, alerts and notification based on the locations of smart vehicles and infrastructure are critical and provide motivation to our scenarios. We deployed a group hierarchy as shown in Fig. 6.4 and build our use cases around it. The proof of concept implementation enforced authorization policies and activity control notifications based on the following scenarios:

Deer Threat Alerts Smart devices and infrastructure which are deployed across geographical area have the ability to detect the area around it and raise alerts for the corresponding groups when an activity or any change is detected. For our scenario, a deer is detected by the motion sensors deployed that trigger the change of the attribute Deer_Threat in location group to value ON. This change pushes alerts to all the current member vehicles of the group. Similar scenarios can also be envisioned for other cases such as speed limit alert, flash floods, accidents, slippery roads, gas station advertisements, or parking garage alerts etc.

Car Pooling Notifications In this scenario, some customer requests a ride to Location-A with his mobile device. The car-pooling request is sent to the vehicles which are near to the customer and have the destination location same as what is requested by the customer. AWS cloud receives the request that calculates the

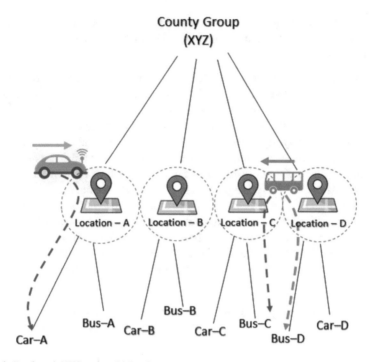

Fig. 6.4 Deployed AWS groups hierarchy

location and relevant groups depending on the location GPS coordinates of the requesting customer so as to raise notification for the cars in the vicinity. In the worst scenario, all the members of the groups Car-A,B,C or D may be sent the request, however, some users would not want to be associated with the car-pooling service or have other criteria such as the rating of the customer to accept or reject the request. The privacy preferences and personal policies of the users should also be evaluated before a notification is sent to the drivers.

6.4.2   Prototype Implementation

Our model is implemented in Amazon Web Services (AWS) for the aforementioned scenarios which are deployed in two phases. The first phase includes the administrative phase that covers the defining of groups hierarchy, the assignment of vehicles into groups as they move and based on their attributes. Further, the inheritance of attributes in the child groups from parent groups and to the members of different groups is also part of the administration. On the other hand, operational aspect involves the deployment of attributes-based policies and dynamic groups which are used to control the activities based on the user and system wide policies, covering the multiple layers of policies needed to make an activity control decision. Policy decision point (PDP) and policy enforcement point (PEP) are created and deployed in our external policy evaluation engine that is attached to AWS to support ABAC authorization.

Administrative Phase A group hierarchy is deployed in Amazon Web Services as illustrated in Fig. 6.4. Four groups are divided based on the geographic location such as Location-A, B, C and D without overlapping area and having a common County-XYZ as the parent group. These location groups can be further divided into Car and Bus sub groups signifying different type of vehicles such as car or a bus. In addition, 10 simulated vehicles were created and their movement is emulated using a python script that continuously publishes MQTT message of the GPS coordinated to the virtual objects or shadows of these vehicle. These GPS coordinates were generated with a Google API as shown by the green dots in Fig. 6.5. As can be seen, the area is divided into four locations and vehicles moving can become member of any subgroup based on their location. Suppose the current location of vehicle Vehicle-1 is in Location-D, and the vehicle pushes an MQTT message with the following payload to a defined topic $aws/things/Vehicle-1/shadow/update in AWS.

```
{"state": {"reported": {"Latitude": "29.4769353",
                        "Longitude":"-98.5018237"}}}
```

This messages reflect that the location now changes to Location-A and together with the vehicle type which is a car in this case, the vehicle is now assigned to the subgroup Car-A within Location-A as reflected in Fig. 6.6. Vehicle type along with the present location to assign them to different location groups already created in the system. We demonstrated these added stand alone services using the Lambda[3] service and the Python based Boto[4] AWS SDK. For the deer threat alert scenario we emulated a location sensor with the ability to detect a deer in its surrounding, that will trigger the change of 'Deer_Threat' attribute for the location group to 'ON' or 'OFF'. This change in the attribute value will further send alerts to all the current members of the group and all its child groups. Fine grained security policies are defined based on the attributes to check which sensors are authorized to modify the value of 'Deer_Threat' attribute for different location groups. The policies are illustrated in Fig. 6.7 reflect that for making such change to the 'Deer_Threat' attribute, a 'Deer_Threat' operation has been defined that must be checked to ensure that the modifying sensor has ID = '1' and is presently located in the same group (Location-A) that needs to be updated. In case, the sensor moves to another location, it will only be allowed to change the attribute of the corresponding group only. Such security policy is important to ensure that only authorized entities and its locations are important to make any such changes.

---

[3] https://aws.amazon.com/lambda/.

[4] https://aws.amazon.com/sdk-for-python/.

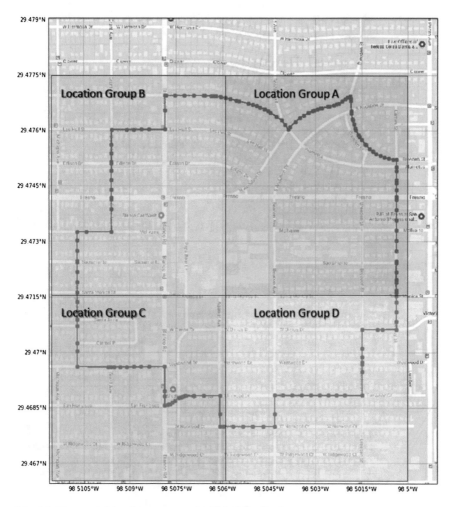

Fig. 6.5 Demarcated location groups and vehicle defined path

The administrative phase of our proposed model in AWS is reflected in Fig. 6.8. When the vehicle moves from one location to another, it keeps on sending its GPS location to the AWS cloud shadow service. Our stand alone service use the location together with the attributes of vehicles and different location groups to determine the membership of vehicle in the pre-defined groups. In case the defined security policy is satisfied with the attribute, the vehicle and group are sent a notification, which will result in inheritance of attributes to vehicles from its new group. As such, when the attribute 'Deer_Threat' is authorized for modification from the location sensor, all the member are correspondingly notified. update_thing_group and update_thing methods are used to support the inheritance of attributes to child groups from parent groups. For our deployment, attributes are inherited to all sub-

```
('Received new coordinates from:', 'Vehicle-1')
Sun May 27 02:56:30 2018
  Location A
     Car-A : [u'Vehicle-1', u'Vehicle-2']
     Bus-A : []
  Location B
     Car-B : []
     Bus-B : [u'Vehicle-6']
  Location C
     Car-C : [u'Vehicle-3', u'Vehicle-4']
     Bus-C : []
  Location D
     Car-D : []
     Bus-D : [u'Vehicle-5']
```

Fig. 6.6 Vehicles assigned to groups dynamically (This is a snapshot at a particular time)

```
{
  "Deer_Threat": {          ◄─────  Policy Operation
    "Source": {
      "1": {          ◄─ ─ ─ ─ ─ ·   Source Attributes
        "Location": {
          "Location-A": {"Group": ["Location-A"]},
          "Location-B": {"Group": ["Location-B"]}
        }
      },
    }
  },                                                        Object Attributes
  "car_pool_notification": {   ◄─────  Policy Operation
    "Source": {
      "Location-A": {    ◄─ ─ ─  Source Attributes
        "destination": {
          "Location-A": {"Notification": ["Car-A"]},
          "Location-B": {"Notification": ["Car-A", "Car-B", "Car-C"]},
          "Location-C": {"Notification": ["Car-C", "Car-D"]},
          "Location-D": {"Notification": ["Car-A", "Car-C", "Car-D"]}
        }
      },
    }
}
```

Fig. 6.7 Fine grained ABAC security policies

groups Car-A and Bus-A of Location-A parent group, and then to all the members in Car-A and Bus-A. Hence, in group Location-A when the attribute 'Deer_Threat' is changed to ON, its modified attributes using Boto describe_thing_group command are:

```
{'Center-Latitude': '29.4745', 'Center-Longitude':
          '-98.503','Deer_Threat': 'ON'}
```

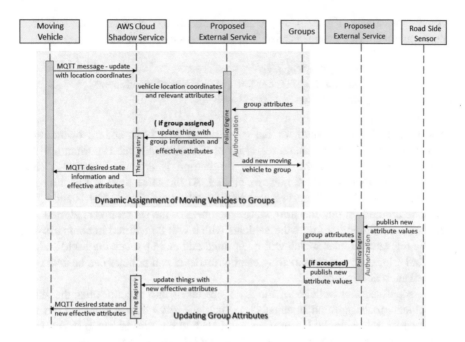

Fig. 6.8 Sequence diagram reflecting the administrative flow

This inherits the attributes to Car-A child group whose effective attributes will be:

```
{'Center-Latitude': '29.4745', 'Center-Longitude':
'-98.503','Deer_Threat': 'ON', 'Location': 'A'}
```

As shown in Fig. 6.6, both Vehicle-1 and Vehicle-2 as member of Car-A, the effective attributes of Vehicle-2 are:

```
{'Center-Latitude': '29.4745', 'Center-Longitude':
'-98.503','Deer_Threat': 'ON', 'Location': 'A',
'Type': 'Car', 'VIN': '9246572903752', 'thingName':
'Vehicle-2'}
```

Operational Phase The operational phase ensures the restrictions imposed due to the security policies created during the administrative phase, which limits the notification and alert activities and supporting multi-level policies together with user personal preferences. For the car pooling scenario, policies are defined to limit alerts for a subset of vehicles within location and other attributes. A requestor is simulated in AWS which needs car pool. This requestor has 'destination' attribute that can be assigned values in Location-A, B, C or D. The requestor creates an MQTT message and sends its current location along with the destination to the

AWS topic $aws/things/Requestor/shadow/update. These attributes determine which subgroups needed to send the service alert from the user.

```
{"state": {"reported": {"policy":
"car_pool_notification", "source": "Location-A",
"destination": "Location-B"}}}
```

The security policy defined for the car_pool_notification operation as shown in Fig. 6.7 specify that a requestor at 'Location-A' (this is current attribute value of location), and the requested destination is around 'Location-A' then members in subgroup Car-A must be alerted. At the same time, if Location-B is the attribute for destination, then members in subgroups Car-A, Car-B and Car-C must be notified. In this scenario, all the members of the subgroup are alerted. The defined security policy limits the vehicles which will be notified in comparison to all the relevant vehicles which will be notified and end up receiving useless alerts. Similarly, location-based marketing can be restricted and policies can be defined to control such notifications.

Personal security policies of the users are taken into account after the subset of vehicles to be notified is calculated. Such security policies subsume the user preferences, for example, if a user does not want to have restaurant notification, or had bad experience with the requesting user and do not want to take him again. In such cases, the alerts will not be notified on the system. Such local user policies can be implemented using the AWS Greengrass service which supports the deployment of local lambda functions on the smart object such as vehicles to provide edge computing ability. This is particularly important for application requiring real time decision making ability and locally enforce personal preferences. An SNS[5] (Simple Notification Service) can be sent to the requestor once the vehicle accepts its request. Figure 6.9 illustrate the sequence of step in case of a car-pooling activity that support multilayer security policies in addition to user privacy preferences.

6.4.3 Performance Evaluation

The performance of the CV-ABAC$_G$ model is evaluated in AWS using various performance metrics when no security policy is evaluated as compared to when our deployed ABAC policies are enforced in the two use cases of car-pool request alerts. Figure 6.10 shows the performance evaluation reflecting the average time (calculated in milliseconds) for the developed policy evaluation engine to evaluate car-pool service request and calculating the subset of vehicles which will be notified. The subset of vehicles reflect scoping of the notifications to only one car which was in the vicinity of the service request as compared to all the associated cars in the

---

[5]https://aws.amazon.com/sns/.

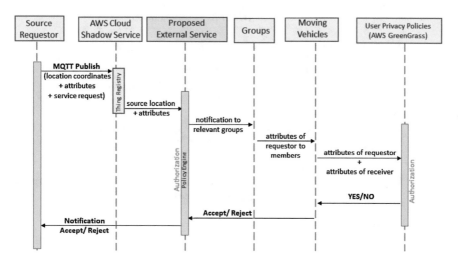

Fig. 6.9 Sequence diagram for the operational phase

Number of Requests	Policy Enforcer Execution Time (in ms)
10	0.0501
20	0.1011
30	0.1264
40	0.1630
50	0.1999

	Cars Notified	
nth Request	With ABAC Policy	Without Policy
41st	2	5
42nd	3	5
43rd	5	5
44th	3	5
45th	2	5
46th	3	5

Fig. 6.10 Policy enforcement time and scoping

ecosystem which may be 100 of miles away from the source of the request. This performance evaluation graph in Fig. 6.11 shows the comparison when no policy is executed (shown by a red line) as compared to the ABAC policy which was reflected in blue. In the experimental proof of concept, as each activity request is similar, a linear graph is achieved with the number of requests increasing since the policy evaluation also increase and therefore the evaluation time. Minor variation in red line is seen due to the latency in the network to reach to the AWS cloud, which is dynamic based on the underlying communication technology used. As can be noted, this external policy evaluation engine has impact on the overall performance of the ITS ecosystem. But, when this proof of concept will be implemented in wide geographic area, this evaluation time will be subsumed by the individual notification time for all the vehicles as compared to only a subset of vehicles will be notified.

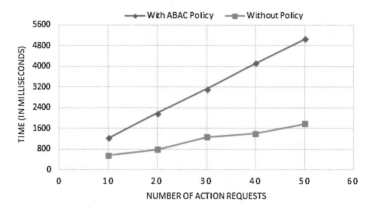

Fig. 6.11 Performance graph

6.5   Summary

This chapter discuss a formal attribute based security model with dynamic groups that consider location sensitive and time focused intelligent transportation. The discussed model proposed dynamic groups which are randomly assigned to different participating entities based on their location and other attributes to support easy management and notification of alerts and services. In addition, the model supports user privacy preferences and have multi-level security policies which are evaluated to make activity access control decision. The implementation of the model is performed in AWS service along with detailed performance analysis.

References

1. Sandhu, R. S., & Samarati, P. (1994).   Access control: principle and practice.   *IEEE Communications Magazine, 32*(9), 40–48.
2. Sandhu, R. S. (1998). Role-based access control. In *Advances in Computers* (vol. 46, pp. 237–286). Elsevier.
3. Ferraiolo, D. F., Sandhu, R., Gavrila, S., Kuhn, D. R., & Chandramouli, R. (2001). Proposed nist standard for role-based access control. *ACM Transactions on Information and System Security (TISSEC), 4*(3), 224–274.
4. Jin, X., Krishnan, R., & Sandhu, R. (2012). A unified attribute-based access control model covering dac, mac and rbac. In *IFIP Annual Conference on Data and Applications Security and Privacy* (pp. 41–55). Springer.
5. Gupta, M., & Sandhu, R. (2016).   The GURA$_G$ administrative model for user and group attribute assignment. In *International Conference on Network and System Security* (pp. 318–332). Springer.
6. Gupta, M., Patwa, F., & Sandhu, R. (2017b). Access control model for the hadoop ecosystem. In *Proceedings of the 22nd ACM on Symposium on Access Control Models and Technologies* (pp. 125–127). ACM.

7. Gupta, M., Patwa, F., Benson, J., & Sandhu, R. (2017a). Multi-layer authorization framework for a representative hadoop ecosystem deployment. In *Proceedings of the 22nd ACM on Symposium on Access Control Models and Technologies* (pp. 183–190). ACM.

8. Gupta, M., Patwa, F., & Sandhu, R. (2017c). Object-tagged RBAC model for the hadoop ecosystem. In *31st Annual IFIP WG 11.3 Conference on Data and Applications Security and Privacy (DBSec)* (vol. 10359, pp. 63–81). Springer Lecture Notes in Computer Science.

9. Gupta, M., Patwa, F., & Sandhu, R. (2018). An attribute-based access control model for secure big data processing in Hadoop ecosystem. In *Proceedings of the Third ACM Workshop on Attribute-Based Access Control* (pp. 13–24).

10. Gupta, M., & Sandhu, R. (2018). Authorization framework for secure cloud assisted connected cars and vehicular internet of things. In *Proceedings of the 23nd ACM on Symposium on Access Control Models and Technologies* (pp. 193–204).

11. Gupta, M., Benson, J., Patwa, F., & Sandhu, R. (2019). Dynamic groups and attribute-based access control for next-generation smart cars. In *Proceedings of the Ninth ACM Conference on Data and Application Security and Privacy* (pp. 61–72).

12. Gupta, M., Benson, J., Patwa, F., & Sandhu, R. (2020). Secure V2V and V2I communication in intelligent transportation using cloudlets. *IEEE Transactions on Services Computing.*

13. Gupta, M., Awaysheh, F. M., Benson, J., Alazab, M., Patwa, F., & Sandhu, R. (2020). An attribute-based access control for cloud enabled industrial smart vehicles. *IEEE Transactions on Industrial Informatics, 17*(6), 4288–4297.

14. Ebert, J., Newton, O., O'Rear, J., Riley, S., Park, J., & Gupta, M. (2021). Leveraging aviation risk models to combat cybersecurity threats in vehicular networks. In *Information, 12*(10). Multidisciplinary Digital Publishing.

Chapter 7
Fine Grained Communication Control for IoT and CPS

7.1 Introduction

Internet of Things (IoT), with "anything" and "everything" connected to the Internet, is a pervasive reality of our lives today. IoT devices are rapidly increasing both in terms of numbers and capabilities. With advancements of IoT enabling technologies, such as Cloud and Edge computing, Artificial Intelligence (AI), and Machine Learning (ML), smart devices are collecting, sharing, and analyzing a huge amount of data associated with the users. For instance, in smart home, there are many smart devices and appliances that are continuously communicating with each other and are collecting data about user behavior patterns. An example scenario is when a user can remotely do the following actions: turn on the thermostat with desired set temperature, play desired music, and set smart cooker to turn on at a set time, so as when the user reaches home, everything is set and user have a relaxing ambiance with cooked food. Some of these IoT devices are already in the market, such as a smart thermostat by NEST [1], or a smart watch that can monitor a user's health and fitness, e.g., Fitbit [2], Apple watch [3].

However, this connected IoT ecosystem introduces many challenges and concerns associated with a huge number of IoT devices that utilize cloud and edge computing services. In the near future, it will be a nightmare for users to manage these billions of connected devices and the data collected from these devices which is stored on specific cloud platforms. With an expected number of 25 billion smart devices by 2025 [4], the IoT attack surface is tremendously expanding. It is a challenge to address security and privacy issues in dynamic and diverse IoT space which includes heterogeneous connected devices, communication protocols and platforms. Thus, a systematic and detailed research approach is essential to secure authorization, communication and data flow in CE-IoT.

7.1.1 Motivation

IoT devices have specific characteristics that make them distinct compared to other connected devices, such as computers, laptops, smartphones, etc. Some of these characteristics are presented below.

- **Distributed and Remote Location:** Today, IoT devices are of different types, such as static and mobile devices, and are widely distributed in large geographical region. These devices are often remotely located with no physical security, unlike personal laptops and desktops.
- **Diverse Nature:** IoT devices are diverse in nature. They are of different sizes and have different capabilities and functionalities. In addition, there are various IoT communication and networking mechanisms or protocols (e.g., MQTT, CoAP), and they are manufactured by different vendors. Also, there are multiple cloud computing platforms that provide customized authentication, authorization, and communication mechanisms.
- **Autonomicity:** Smart ecosystems are envisioned to be autonomous environments where devices can interact with each other and perform operations automatically. Thus, smart devices and applications in any IoT use case must be capable of acting autonomously utilizing technologies like AI and ML along with cloud services, i.e., storage, computation, analytics.
- **Dynamic Behavior**: IoT comprises various application domains, such as smart homes, smart health, smart transportation, etc. The sensors and devices in each application domain may be different and even same sensors or devices can behave differently based on the use case scenarios in any application domain. Based on the characteristics and contexts for different users, the devices may collect different types of data. Therefore, use case scenario, application domain, and contextual parameters play a critical role in how these devices behave and can be secured and managed effectively.

With these evolving characteristics, managing security and privacy in IoT ecosystem has become even a bigger challenge. This chapter mainly focuses on access and communication control aspects of security in CE-IoT. With emerging access control requirements of futuristic IoT applications, traditional access control models are not sufficient and there is a need for a novel access and communication control framework. Currently, most IoT access control models [4, 32–36] have focused on a single centralized cloud IoT platform and utilized the most dominant access control model, viz Role-Based Access Control (RBAC) [5, 6]. However, IoT ecosystem is evolving and moving beyond a single cloud platform. Similarly, in the future, the authorizations in IoT will be managed or shared across a set of collaborating cloud providers or servers and will become decentralized [4]. In the context of a diverse and dynamic IoT environment, it is inevitable to consider different characteristics (or attributes) of users, devices, and context beyond *roles* in identifying the authorizations associated with IoT devices and applications.

Besides access control, communications and data flow in CE-IoT architecture need to be secured against unauthorized data access and modifications. With ubiquitous IoT devices collecting, storing, and sharing sensitive user data, there is a critical need to develop a novel communication and data flow control mechanism for securing data at rest (stored data) and data flow (data in-transit), which can also preserve user data privacy. However, an access and communication control framework for IoT still hasn't been developed. This chapter presents an attribute-based access control and communication control framework, known as **ABAC-CC**. This framework is designed to secure accesses and communications in CE-IoT architecture which includes several IoT devices, gateways, and cloud services. ABAC-CC utilizes attributes of different entities, such as users, devices, gateways, etc. for determining if specific communications and data flow between various IoT entities should be allowed or denied.

This chapter also introduces a novel **Attribute-Based Communication Control (ABCC)** model that secures data communication and flow between several components in cloud and edge network based on attribute-based communication control policies. These policies are defined using attributes of relevant IoT entities, such as devices, gateways, and virtual objects (VOs) and a new type of attribute, i.e., the **message attributes**. The ABCC model incorporates this novel attribute, *message attributes*, which are derived from the content or data of the message itself. IoT devices are continuously collecting and sharing messages with other entities in the architecture. A message is the *unit of communication* between IoT devices. The communication control policies utilize message/data attributes, *(attribute name and value pairs)* along with attributes of other entities, including users, devices, and gateways [7]. Moreover, based on user's privacy concerns, users can define communication control policies utilizing the attributes of various entities in the system. With future advancements in IoT, we identify ABCC model as an essential component of the ABAC-CC framework which will enable secure communications and data flow across smart devices, gateways, and different cloud platforms.

7.1.2 Chapter Organization

This chapter is organized into five sections. We first discuss brief background and related work on IoT access and communication control models in Sect. 7.2. In Sect. 7.3, we discuss access control and communication control requirements in dynamic IoT ecosystem and some use case scenarios. In Sect. 7.4, we present a basic conceptual ABCC model and its entities and types of attributes that are used in defining ABCC policies. Finally, the ABAC-CC framework is presented in Sect. 7.5 along with its applicability analysis based on specific use case scenarios. To conclude, summary is provided at the end of the chapter.

7.2 Background and Related Work

This section provides a brief background on different types of CE-IoT paradigms and discusses relevant prior research on attribute-based access and communication control mechanisms.

7.2.1 CE-IoT Architectures

Generally, a basic IoT architecture comprises three layers: *(i) Object or perception layer*, including physical objects, *(ii) One or more Middleware layer(s)*, that include virtual objects (digital representation of physical objects) [8], and Service-Oriented Architecture (SOA) management services, and *(iii) An Application layer*, the top layer of the architecture which allows users and administrators to directly interact with these applications. As discussed earlier, several layered IoT architectures have been proposed in the literature [9–13]. An access control oriented (ACO) architecture for CE-IoT is proposed in [14] by Alshehri and Sandhu. It has four layers: *an object layer, virtual object layer, cloud services layer,* and *applications layer* and each layer encapsulates different cloud and IoT entities, associated data, and their access control requirements. To abstract the heterogeneity of IoT devices, and enable edge computing capabilities, particularly in domains like Wearable IoT (WIoT), Bhatt et al. [15] enhanced the ACO architecture and proposed an Enhanced ACO architecture (EACO). Figure 7.1 shows the two layered CE-IoT architectures. The ABAC-CC framework is based on the EACO architecture and focuses on access and communications between IoT entities in EACO architecture.

7.2.2 Related Work

IoT data is mainly two types: *(i) static data or data at rest*, and *(ii) dynamic data or data in motion*. Current access control models focus on securing static data stored on machines. However, IoT sensors, devices, gateways, virtual objects (VOs), and cloud services are continuously communicating and sharing data with each other in CE-IoT and the data communication and data flow need to be secured. There is extensive literature on access control models, but there is very limited research on communication control models. Some prior research work have developed access control models to secure access to data in databases [16, 17] using role-based approach and some special attributes. Generally, communication control has been significantly studied in the networking domain. In computer networks, there are distinct networking devices and systems, such as routers and firewalls. These devices allow controlling communication and network traffic flow in form of packets

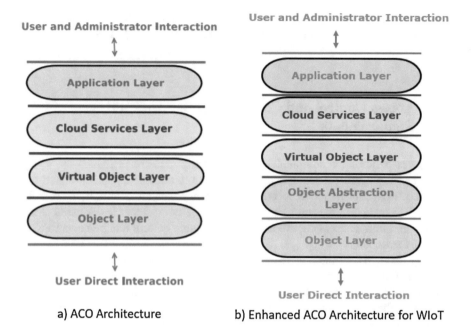

Fig. 7.1 Access control oriented (ACO) and enhanced ACO CE-IoT architectures

by specifying specific rules. An example of a communication control device or system is a **Guard** device. *Guards* control data communication from one component to the other in a computer network [7]. Inspired by these guards, we developed a conceptual model of ABCC. This is the first conceptual model of ABCC to the best of our knowledge. In the future, we believe that formal communication control models based on attributes of IoT entities can be designed and developed to secure communications among various authorized entities in the CE-IoT architecture.

7.2.3 Scope and Assumptions

We mainly focus on current and evolving access control and communication control requirements in CE-IoT. Access control is a well-understood and discussed concept, whereas communication control is a novel concept that needs to be explored in detail. Nonetheless, Fig. 7.1 shows that the **Device-to-Device** communication occur through the *Object Abstraction layer*. In some critical IoT domains, such as Vehicular IoT and Internet of Battlefield or Military Things (IoMT/IoBT), device-to-device communications are critical and privacy-sensitive and are enabled by edge computing architecture. In this chapter, we assume authentication of physical devices is enabled through cryptographic keys and certificates. Hence, discussions associated with device-to-device communications at the object layer are out of scope.

7.3 Access Control and Communication Control Requirements

The access control and communication control requirements in CE-IoT are recognized by investigating existing boundaries, dissimilarities, and gaps in the CE-IoT architecture to improve an access control and communication control framework for CE-IoT. In this part, relevant issues within the IoT ecosystem are discussed. Also, current and future access control and communication control requirements for CE-IoT are explored.

Nowadays, most common CE-IoT architecture is one Cloud-IoT architecture, where IoT devices interconnect with a single Cloud platform. In order to assist billions of IoT devices interactions and offer local computation, analytics, and storage at the edge of the network, a CE-IoT architecture with edge computing is illustrated through little edge cloudlets [15, 18]. These cloudlets could be implemented as gateways, that have adequate capabilities like storage and computation power, to serve as a small cloud on the edge of the network. On the other hand, influenced by assorted instances of the Cloud-IoT architecture, the access control and communication control needs will grow upon the architecture employed for IoT devices and applications.

As previously discussed, there is no formal CE-IoT architecture for IoT yet. Depending upon various scenarios, users can adjust different instances of the CE-IoT architecture. It is partially due to the present marketing strategy of CSPs to improve and deploy their IoT devices, for example, several compatible Smart Home Assistants can only interconnect with the Cloud platform that manufactured it. Consequently, a single centralized Cloud-IoT architecture with billions of IoT devices raises an **interoperability** concern, which is being recognized by customers who possess many smart devices from multiple producers or vendors. Thus, a centralized cloud-IoT architecture needs to be evolved with real-time communications and cooperation through multiple Cloud platforms [4]. With inter-cloud cooperation being unavoidable in the future, effective and elastic access control and communication control mechanisms are needed to allow partnerships and trust among single-cloud and multi-cloud platforms.

Two different examples of the Cloud-Enabled IoT architecture are illustrated in Fig. 7.2. A Cloud-IoT architecture without edge computing in line with the ACO architecture is presented in Fig. 7.1a. A Cloud-IoT architecture with edge cloudlets, which permit edge computation, communication, and storage in line with the ACO architecture is presented in Fig. 7.1b. The CE-IoT architecture in Fig. 7.2b is proper to assist local computation and analysis near the edge by applying AI

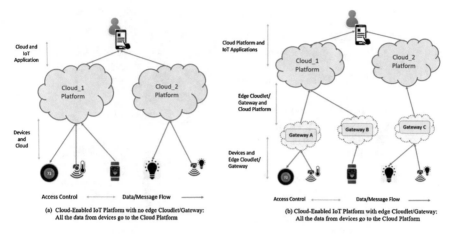

Fig. 7.2 Access and communication control shown in different CE-IoT architectures

and ML techniques and assist real-time communications with rapid reply even in irregular network once the gateway (cloudlet) and devices are authenticated and configured across the Cloud. In the two architectures (a) and (b), the access control and communication control needs will vary upon on various engaged entities and user privacy issues. For instance, in (a) architecture, IoT devices associate with the Cloud and direct all collected data to the Cloud. In the cloud platform, the devices will communicate with their virtual objects (VOs), which can be accessed by IoT applications and other services to get, send, or update data on physical devices.

On the other hand, the architecture in Fig. 7.2b, devices associate and communicate with the edge gateway/cloudlet, which offer edge computation and permits access control and communication control near the edge of the network. The gateway also allowed communication with the VOs in the cloud and assure that physical devices interconnect with analogous VOs in the Cloud platform. Furthermore, the edge cloudlets allow users to outline privacy preserving communication control policies. For instance, users with greater privacy worries do not wish to transfer their data to the cloud platform at all, but they prefer to keep their data within the edge network, maybe kept in the edge cloudlet (gateway) except if certain emergency or unexpected condition happens.

Figure 7.3 displays a complete view of CE-IoT with various entities like users, devices, gateways in cloudlets, and virtual objects in the Cloud platform as well as other services. The architecture in this figure is a detailed version of Fig. 7.2b. The figure describes a Smart Health scenario, where users have various wearable IoT devices that connect to gateways, which are connected to the Cloud. Cloud-Enabled IoT platforms like AWS IoT and Google IoT Core use certain adapted form of RBAC models to manage authorization and access control. An Identity and Access Management (IAM) service in the Cloud displayed as ACP (access control policies). Cloud-Enabled IoT platforms have recognized the limitations of RBAC,

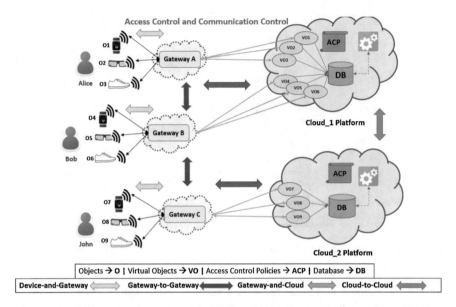

Fig. 7.3 Access control and communication control requirements in CE-IoT

so they have started looking into more powerful models, such as ABAC, which has not been successfully implemented yet.

Simultaneously, the flow of data from one point to another one should be controlled. For example, the flow of data from devices to gateway and then to VO in the Cloud platform needs to be controlled. Also, the flow of data between gateways and between many Cloud platforms. Figure 7.3 shows the possible accesses and communications in different colors as **Gateway-to-Gateway** (access and communication across gateway and gateway), **Device-and-Gateway** (access and communication between device and gateway as well as gateway and device), **Gateway-and-Cloud** (access and communication between gateway and VO in the Cloud platform as well as from VO in the Cloud platform and gateway), and **Cloud-to-Cloud** (access and communication across VOs in various Cloud platforms).

7.3.1   Use Case Scenarios

Based on Fig. 7.3, use case scenarios are explained in the following part.

- **Scenario 1:** In the smart health monitoring scenario in Fig. 7.3, users do not desire to share their data with the cloud at all, and rather they would prefer to restrict their data at the edge network and send necessary updates to the cloud platform upon certain predefined situations.

- **Scenario 2:** Likewise, users desire to limit incoming data from cloud to users through IoT applications, such as recommendations for health and exercise, which are less important inherently.

Within these scenarios, there are several questions that need to be answered. For example, *How would a user be able to control communications in these scenarios? What is a secure and flexible way to do so? How would they define access control policies together with communication control policies?* These are some of the specific questions that require further research and can be facilitated through the ABAC-CC framework. The ultimate goal is to enable the users to have the flexibility to define fine grained access control and communication control policies for their smart devices. A promising approach is to utilize the **attribute-based** approach for access and communication control within the ABAC-CC framework.

7.4 Attribute-Based Communication Control

This section presents a general conceptual model for Attribute-Based Communication Control (ABCC). The ABCC model is then compared with the Attribute-Based Access Control (ABAC) and its components. Mostly, access control implies controlling access (e.g., read, write) on a specific protected entity, which could be an object, or a subject, from another entity, such as a user or a subject, who is requesting access on it. Whereas, in communication control, the communication of a specific element, i.e. the message is controlled from one entity or endpoint to other entity or endpoint. Depending upon the IoT application domain or use case scenario, specific entities and their characteristics in the ABCC model can be defined more concretely for real-world implementation purposes.

7.4.1 Attribute-Based Access Control Model

Recently, ABAC is receiving significant attention in the IoT domain mainly because of its dynamic and flexible nature. In the literature, many ABAC models have been proposed and developed for different application domains [19–26]. Besides operational context, ABAC models have also been applied in administrative context for securing administrators' accesses, which mainly includes CRUD (create, delete, update, revoke) operations on model entities—users, objects, subjects, roles, and virtual objects [27–30]. Figure 7.4a presents a simplified ABAC model with its various components. It includes subjects (S), objects(O), subject attributes (SA), objects attributes (OA) and operations (OP). A subject is a logical entity (e.g., a process) or an individual user. An object is a resource (e.g., printer, file) or data stored in a system. An operation is an access right, such as read, write, credit, debit operations, that need to be performed on a specific object by a subject.

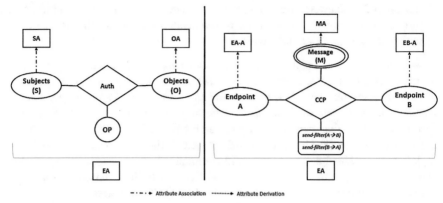

a) Attribute-Based Access Control (ABAC) Model   b) Attribute-Based Communication Control (ABCC) Model

Fig. 7.4 Attribute-based access control vs. attribute-based communication control

Subjects and objects have attributes which represent their characteristics. Examples of subject attributes are name, age, title, etc., and object attributes are object's owner, type, sensitivity level, etc. In addition, there are contextual or environmental attributes (EA) which are system wide attributes, such as time, location, etc. All these attributes are utilized in defining fine grained access control policies. An example of an attribute-based authorization policy is—*a user with title as manager can read an object with sensitivity level as high when the time of the day is between 9:00 am to 5:00 pm and location is office.* ABAC policies can be specified in two ways: *logical formulas with predicate logic* and *enumerations policy* [31].

7.4.2   *Attribute-Based Communication Control Model*

ABAC model employs attributes of different entities in specifying the access control policies. Whereas ABCC model utilizes both attributes of entities communicating with each other and attributes of the communication unit (message) in communication control policies for determining if the communication and data flow between two components should be allowed or denied. Prior literature have identified the need for controlling data flow and communication in IoT. Alshehri and Sandhu developed access control models to control virtual object (VO) to virtual object (VO) communications [14, 30]. However, a conceptual model for attribute-based communication control has not been developed yet.

Figure 7.4b shows a conceptual model for attribute-based communication control. ABCC model has distinct features and characteristics in comparison to the ABAC model. In ABCC model, there are two endpoints **Endpoint A** and **Endpoint B**, and a **Message**, which is being communicated between two endpoints. Based on different contexts, endpoints could be different types of devices, such as routers (stateless/stateful or internal/external routers), systems, and/or smart connected devices. *Endpoint A* and *Endpoint B* have attributes assigned to them which represent their characteristics and properties, such as *type, owner,* etc. In the conceptual model, attributes of *Endpoint A* and *Endpoint B* are represented as **EA-A** and **EB-A** respectively. The *message* is a unique new element which is generated and comes into existence when an endpoint gathers data and generate a message with that data, and then an endpoint can send it to, and/or receive it from another endpoint during the process of communication and data flow. The message would have a definite structure, such as JSON, XML, and have a set of properties in it. The message attributes **MA** and their values are derived from the properties within a message and are not assigned by an administrator. The properties in the message content, arranged as key and value(s) pairs, are essentially derived as the message attributes based on a use case scenario. For instance, a message has a property as *temp = 80* where temp is the *key* and 80 is the *value*, and we can derive it as a message attribute *temp* with a value 80.

In ABCC model, there is a directional one-way operation **send-filter** which allows to send a message from a sender to a receiver. The send-filter operation has two instances representing two directions of communication (Endpoint A to Endpoint B, and Endpoint B to Endpoint A). Suppose there are two entities devices and cloud, so there can be two communication scenarios, such as an IoT device (sender) is sending messages to cloud (receiver), and cloud (sender) sending messages to devices (receiver). These two instances are distinct since there are different communication control and information flow requirements in the two directions. The *send-filter($A \rightarrow B$)* represents a *send-filter* operation where a message is being sent from *Endpoint A* to *Endpoint B*. Similarly, *send-filter($B \rightarrow A$)* represents that the message is being sent from *Endpoint B* to *Endpoint A*.

The send-filter function is defined with 2 inputs, a *sender* and a *receiver*, and is evaluated based on an attribute-based communication control policy. This policy is specified in terms of sender attributes, receiver attributes, message attributes and environment attributes. The communication control policy function (CCP) evaluates the communication request based on the policy and relevant attributes and determines the decision for communication and data flow as shown in Fig. 7.4b). There are three possible results of the evaluation– i) message is blocked, ii) original message is forwarded, or iii) a sanitized message is forwarded with sensitive portions removed from the message. For instance, let us assume that there is an "owner" attribute for two endpoints (Endpoint A and Endpoint B), and we have the following examples of communication control policies.

- *If the owner attribute values for a gateway, which is endpoint A, and owner attribute values for a virtual wearable IoT device, which is endpoint B, are the same, and there is a message attribute, temperature value whose value is greater than 102 degrees Fahrenheit, then the send-filter operation would evaluate to sending an unfiltered or original message from endpoint A to B.*
- *For the above policy, if temperature value is within a normal range, then send-filter operation would either evaluate to sending a filtered message removing*

sensitive piece of information that is location of the user, or may even evaluate to not sending the message to cloud and store it at gateway (endpoint A) since temperature value is within a normal range.

In order to secure communication and data flow, a set of communication control policies are defined by a user or an administrator based on the attributes of endpoints and messages in a system. For a specific sender, receiver (target), and a message, the **Communication Control Policy (CCP)** function is evaluated to identify if the message should be sent unfiltered (original message), filtered (removing sensitive information), or should not be sent from a sender to a receiver. As per the direction of communication and data flow, either of the endpoints can act as a sender or a receiver of a message. **Environment attributes (EA)** can also be included in communication control policies for enabling more fine grained and dynamic communication control based on a specific context, such as time of day, location, etc. A simple communication control policy is given as: *"if the owner of Endpoint A and Endpoint B is the same, then allow the message to be sent from A to B, otherwise deny."* In ABCC model, CCP function could be co-located with any one of the two endpoints, and in some cases, it can also be deployed in a separate system between two endpoints. In CE-IoT architecture, several components are interacting with each other and a large amount of data is continuously flowing between several components in a system. For example, in a WIoT scenario, messages are communicated between wearable devices, gateways, virtual objects (VOs), cloud services, and applications. Hence, in the model, endpoints, their attributes, messages, and direction of communication and data flow changes and evolves based on the type of communication architecture under consideration.

While there are numerous ABAC models proposed by researchers in different contexts and application domains, this is the first general conceptual ABCC model to the best of our knowledge. It allows to control communication between two endpoints based on their attributes as well as message attributes. The conceptual ABCC model is an abstract model whose entities can be realized into concrete entities and components based on the communication paradigm being used (e.g. publish-subscribe) in a real-world scenario. Today, some of the prevalent IoT communication models are publish/subscribe model, and widely adopted TCP/IP communication model for Internet communications.

Both ABAC and ABCC models employ attributes of different entities in the system, however, the units (messages) being controlled are dynamic and distinct. In addition, the gist of ABCC lies in its use of message attributes together with attributes of other entities in the system. Besides, a major difference is type of data being protected, for instance, ABAC focuses on protecting the data stored in a system, whereas ABCC secures data and information in motion (or in-transit). ABCC model is also unique compared to ABAC since it focuses on two major security concerns. First, it checks if two endpoints should be allowed to communicate with each other utilizing their attributes. Second, it secures communication and data flow from one endpoint to another endpoint while considering the content of data (or messages). This is critical especially to preserve user privacy and data security while data is in transit or is mobile. Another critical difference is that the endpoints are system entities or machines (e.g., devices, gateways) in active states rather than individuals. It is essential to mention that a user's identity and information is embedded in one of the endpoints and its attributes, and the message itself that is being communicated between two endpoints. In summation, while ABCC model and its capabilities are relevant and applicable in numerous application domains, but this paper focuses on ABCC models for CE-IoT architecture.

7.5 Attribute-Based Access and Communication Control Framework

There are several entities and components that continuously interact with each other, such as users, IoT things, devices, gateways, virtual objects, cloud services, and applications. These interactions include authorizations defined for these components and data communications among these components. Cloud computing providers mostly utilize their existing cloud access control architecture to IoT, however, IoT has unique characteristics that distinguish it from cloud, primarily due to distributed and autonomous nature of IoT devices (e.g., sensors, actuators) and gateways deployed in the physical environment. Therefore, it is necessary to reevaluate traditional access control models and develop new models to secure IoT ecosystem. An Attribute-Based Access Control and Communication Control (ABAC-CC) framework is proposed to adequately capture and address evolving access control and communication control requirements in CE-IoT architecture.

7.5.1 ABAC-CC Framework

Figure 7.5 depicts the attribute-based access control and communication control (ABAC-CC) framework across the Enhanced ACO layers. The ABAC-CC framework is based on the attribute-based approach which forms the foundation of the framework to control access to various entities and to control communication in terms of data flow from one end to another in CE-IoT services/applications. The core components of ABAC-CC framework are described as follows.

- **Authentication**: In CE-IoT, physical IoT devices can authenticate using cryptographic key coupling between physical devices and virtual objects (things or device shadows). These keys are generated and managed by the cloud service providers for their own platform and devices connecting to their platforms.
- **Attribute-Based Access Control and Authorization**: For securing the CE-IoT architecture and its components, ABAC models need to be developed and enforced for enabling fine grained access control and authorization policies in CE-IoT. These models can be defined and implemented at the cloud level and enforced on specific entities at lower layers in the EACO architecture.

Fig. 7.5 Attribute-based
access and communication
control framework in EACO
layers

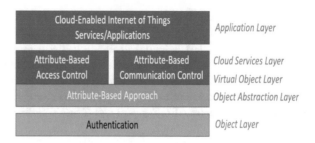

- **Attribute-Based Communication Control**: ABCC policies are defined in terms of entities in a system and message attributes. Furthermore, users can also define their own desired privacy policies based on the access control and communication control needs and requirements. By incorporating the message attributes which are derived from the message itself by inspecting the message content, ABCC adds significant flexibility and enhanced data security and privacy. It allows to control the flow of IoT data/messages from one component to another in the CE-IoT architecture. A critical question is where do we deploy the ABCC model, either cloud or edge of the network. An ideal place to deploy and enforce communication control policies would change and can be adapted as required. If a user wants to secure data flow from edge to cloud, then the user should deploy ABCC policies at the gateway or cloudlet level rather than cloud.
- **Cloud-Enabled IoT Services and Applications**: Applying ABAC-CC framework, which is based on attribute-based approach, CE-IoT applications and services can enable fine grained access and communications compared to their current role-based and policy-based approaches.

We can utilize ABAC-CC framework in controlling data access and communication between different components, *Devices-and-Gateways*, *Gateways-to-Gateways*, *Gateways-and-Cloud(Virtual Objects)*, and *Cloud-to-Cloud*, as shown in Fig. 7.3. We consider a cryptographic key coupling authentication for physical devices for enabling *Devices-to-Devices* communication. Mostly, in edge computing enabled CE-IoT, we consider that device-to-device access and communication is enabled through edge gateways. In the near future, domains like Vehicular IoT and IoBT/IoMT will require edge computation support and the ABAC-CC framework needs to be expanded to incorporate relevant Device-to-Device access and communication.

Figure 7.6 represents a smart health use case scenario. Here, a user has wearable devices that are continuously collecting user data, i.e., their physiological parameters and behavior. It also represents a simple mapping of the ABAC-CC framework and its application in the context of the use case with attribute-based access and communication control policies. In this use case, the user (Alice) wants to restrict her *location* to be sent to the cloud. Other data values, such as *temperature* and *heartrate*, given they are in their normal range, can be stored at the edge gateway or cloud, or else may be its allowed to flow from the gateway to virtual object in the cloud.

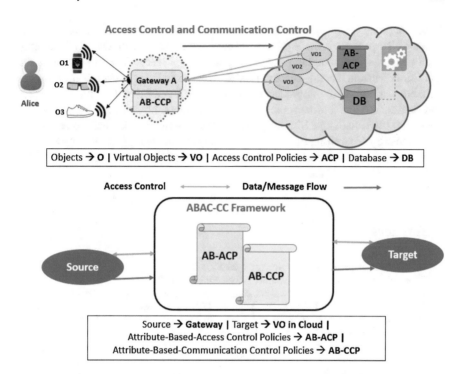

Fig. 7.6 Access and communication control from edge cloudlet to cloud utilizing ABAC-CC framework

In this use case, the framework first evaluates the attribute-based access control policy. For example, if a device and gateway have the same *owner*, then they have access on each other. Next, it evaluates communication control policies, which will be evaluated with entities and message attributes including *heartrate*, *location*, and *temperature* which are message attributes. If we have message attributes with a specific range (normal range) in the policy file and policies are satisfied, then data will be stored on edge gateway. However, more research is needed to answer specific questions, such as *How formal models and definitions will be devised for ABCC models?, How the ABAC and ABCC policies can be defined together in a single or distributed platform?*, and so on. In addition, for today's CE-IoT platforms, there need to be a gradual shift from current RBAC models to the ABAC-CC framework in real-world use case scenarios and implementation.

7.6 Summary

This chapter presented the conceptual Attribute-based Communication Control (ABCC) model and compared it against ABAC model. For securing the data communication and flow between various CE-IoT components, an **Attribute-**

Based approach is proposed while including edge computing capabilities. The chapter also presented the ABAC-CC framework and its applicability in a smart health use case. In summary, the main goal of this research is to reevaluate and reconsider current access control models for CE-IoT and develop new models utilizing attribute-based approach for securing IoT authorizations and data at rest and data in motion. Furthermore, real-world implementation and enforcement of ABAC is still a challenge. Similarly, ABCC is a new model that needs to be adopted in the industry. Therefore, implementation and enforcement of ABAC together with ABCC models are necessary in real-world cloud-IoT platforms, such as AWS IoT and Google Cloud IoT Core.

References

1. Google Nest. https://nest.com/. Accessed: 2020-01-08.
2. Fitbit. https://www.fitbit.com/us/home. Accessed: 2020-01-08.
3. Apple Smart Watch. https://www.apple.com/apple-watch-series-5/. Accessed: 2020-01-08.
4. Tang, B., Kang, H., Fan, J., Li, Q., & Sandhu, R. (2019). Iot passport: A blockchain-based trust framework for collaborative internet-of-things. In *Proceedings of the 24th ACM Symposium on Access Control Models and Technologies* (pp. 83–92).
5. Sandhu, R., Coyne, E. J., Feinstein, H., & Youman, C. (1996). Role-based access control models. *IEEE Computer, 29*(2), 38–47 (1996)
6. Ferraiolo, D. F., Sandhu, R., Gavrila, S., Richard Kuhn, D., & Chandramouli, R. (2001). Proposed NIST standard for role-based access control. *ACM Transactions on Information and System Security (TISSEC), 4*(3), 224–274.
7. Bhatt, S. (2018). *Attribute-Based Access and Communication Control Models for Cloud and Cloud-Enabled Internet of Things*. Ph.D. thesis, University of Texas at San Antonio.
8. Nitti, M., Pilloni, V., Colistra, G., & Atzori, L. (2016). The virtual object as a major element of the internet of things: A survey. *IEEE Communications Surveys & Tutorials, 18*(2), 1228–1240.
9. Al-Fuqaha, A., Guizani, M., Mohammadi, M., Aledhari, M., & Ayyash, M. (2015). Internet of things: A survey on enabling technologies, protocols, and applications. *IEEE Communications Surveys & Tutorials, 17*(4), 2347–2376.
10. Atzori, L., Iera, A., & Morabito, G. (2010). The internet of things: A survey. *Computer Networks, 54*(15), 2787–2805.
11. Porambage, P., Ylianttila, M., Schmitt, C., Kumar, P., Gurtov, A., & Vasilakos, A. V. (2016). The quest for privacy in the internet of things. *IEEE Cloud Computing, 3*(2), 36–45.
12. Yang, Z., Yue, Y., Yang, Y., Peng, Y., Wang, X., & Liu, W. (2011). Study and application on the architecture and key technologies for IoT. In *2011 International Conference on Multimedia Technology* (pp. 747–751). IEEE.
13. Gupta, M., & Sandhu, R. (2018). Authorization framework for secure cloud assisted connected cars and vehicular internet of things. In *Proceedings of the 23nd ACM on Symposium on Access Control Models and Technologies* (pp. 193–204).
14. Alshehri, A., & Sandhu, R. (2016). Access control models for cloud-enabled internet of things: A proposed architecture and research agenda. In *2nd International Conference on Collaboration and Internet Computing (CIC), 2016, IEEE* (pp. 530–538). IEEE.
15. Bhatt, S., Patwa, F., & Sandhu, R. (2017). An access control framework for cloud-enabled wearable internet of things. In *2017 IEEE 3rd International Conference on Collaboration and Internet Computing (CIC)* (pp. 328–338). IEEE.

16. Byun, J.-W., Bertino, E., & Li, N. (2005). Purpose based access control of complex data for privacy protection. In *Proceedings of the tenth ACM Symposium on Access Control Models and Technologies* (pp. 102–110).
17. Rabitti, F., Bertino, E., Kim, W., & Woelk, D. (1991). A model of authorization for next-generation database systems. *ACM Transactions on Database Systems (TODS), 16*(1), 88–131.
18. Satyanarayanan, M., Bahl, P., Caceres, R., & Davies, N. (2009). The case for VM-based cloudlets in mobile computing. *IEEE Pervasive Computing, 8*(4).
19. Yuan, E., & Tong, J. (2005). Attributed based access control (ABAC) for web services. In *IEEE International Conference on Web Services (ICWS05)*. IEEE.
20. Shen, H.-b., & Hong, F. (2006). An attribute-based access control model for web services. In *2006 Seventh International Conference on Parallel and Distributed Computing, Applications and Technologies (PDCAT06)* (pp. 74–79). IEEE.
21. Lang, B., Foster, I., Siebenlist, F., Ananthakrishnan, R., & Freeman, T. (2009). A flexible attribute based access control method for grid computing. *Journal of Grid Computing, 7*(2), 169–180.
22. Bhatt, S., Patwa, F., & Sandhu, R. (2016). An attribute-based access control extension for openstack and its enforcement utilizing the policy machine. In *IEEE 2nd International Conference on Collaboration and Internet Computing (CIC)* (pp. 37–45). IEEE.
23. Bhatt, S., Patwa, F., & Sandhu, R. (2017). Abac with group attributes and attribute hierarchies utilizing the policy machine. In *Proceedings of the 2nd ACM Workshop on Attribute-Based Access Control* (pp. 17–28). ACM.
24. Biswas, P., Sandhu, R., & Krishnan, R. (2016). Label-based access control: An abac model with enumerated authorization policy. In *Proceedings of the 2016 ACM International Workshop on Attribute Based Access Control* (pp. 1–12). ACM.
25. Servos, D., & Osborn, S. L. (2014). Hgabac: Towards a formal model of hierarchical attribute-based access control. In *International Symposium on Foundations and Practice of Security* (pp. 187–204). Springer.
26. Gupta, M., Benson, J., Patwa, F., & Sandhu, R. (2020). Secure V2V and V2I communication in intelligent transportation using cloudlets. *IEEE Transactions on Services Computing*. https://doi.org/10.1109/TSC.2020.3025993.
27. Gupta, M., & Sandhu, R. (2016). The $GURA_G$ administrative model for user and group attribute assignment. In *International Conference on Network and System Security* (pp. 318–332). Springer.
28. Gupta, M., Patwa, F., & Sandhu, R. (2018). An attribute-based access control model for secure big data processing in hadoop ecosystem. In *Proceedings of the Third ACM Workshop on Attribute-Based Access Control* (pp. 13–24). ACM.
29. Ninglekhu, J., & Krishnan, R. (2017). Attribute based administration of role based access control: A detail description. Preprint. arXiv:1706.03171.
30. Alshehri, A., & Sandhu, R. (2017). Access control models for virtual object communication in cloud-enabled IoT. In *International Conference on Information Reuse and Integration (IRI), IEEE* (pp. 16–25). IEEE.
31. Biswas, P., Sandhu, R., & Krishnan, R. (2016). A comparison of logical-formula and enumerated authorization policy ABAC models. In *IFIP Annual Conference on Data and Applications Security and Privacy* (pp. 122–129). Springer.
32. Bhatt, S., Pham, T. K., Gupta, M., Benson, J., Park, J., & Sandhu, R. (2021). Attribute-based access control for AWS internet of things and secure Industries of the Future. *IEEE Access, 9*, 107200–107223.
33. Gupta, M., & Sandhu, R. (2021). Towards activity-centric access control for smart collaborative ecosystems. In *Proceedings of the 26th ACM Symposium on Access Control Models and Technologies* (pp. 155–164).
34. Gupta, M., Awaysheh, F. M., Benson, J., Alazab, M., Patwa, F., & Sandhu, R. (2020). An attribute-based access control for cloud enabled industrial smart vehicles. *IEEE Transactions on Industrial Informatics, 17*(6), 4288–4297.

35. Gupta, M., Patwa, F., Benson, J., & Sandhu, R. (2017). Multi-layer authorization framework for a representative Hadoop ecosystem deployment. In *Proceedings of the 22nd ACM on Symposium on Access Control Models and Technologies* (pp. 183–190).
36. Gupta, M., Patwa, F., & Sandhu, R. (2017). Object-tagged RBAC model for the Hadoop ecosystem. In *IFIP Annual Conference on Data and Applications Security and Privacy* (pp. 63–81). Springer.

Chapter 8
Conclusions and Future Work

8.1 Summary

This section summarizes the main contributions and conclusions presented in this book.

8.1.1 IoT and CPS Access Control Oriented Architectures

In this brief, we first focused on understanding the needs of access control and authorization solution in IoT and CPS domains. After we motivated our case, we discussed access control oriented architecture (ACO) having four layers including object layer, virtual object layer, cloud layer, and application layer. Each of these layers encapsulate different entities, associated data, and their access control requirements in the framework. In addition, to support the gateway needed for constrained IoT devices such as medical and wearable IoT, we discussed the enhanced ACO architecture with an additional Object Abstraction layer. We then focused on intelligent transportation systems due to its dynamic and mobile nature unique to other CPS systems. We presented the extended access control oriented architecture (E-ACO), which extends the ACO architecture with the introduction of clustered objects. These clustered objects reflect the smart objects which have multiple sensors, similar to smart cars, having 100s sensors inside it having different functionality. Further, it also reflects possible interactions between sensors in same clustered object or between different object's sensors. The architectures discussed are used in the later part of this book as a reference to build access control models for cloud and edge supported Internet of Things and Cyber Physical Systems. We also discussed use cases specifically catered with these different access control architectures.

M. Gupta et al., *Access Control Models and Architectures For IoT and Cyber Physical Systems*, https://doi.org/10.1007/978-3-030-81089-4_8

8.1.2   Authorization Frameworks

We next defined different access control frameworks that reflect authorization needs at various layers of extended and enhanced ACO architecture. We adapted the CE-IoT architecture from a WIoT context and referred to it as the Cloud-Enabled Wearable Internet of Things (CE-WIoT). We presented the five-layered Enhanced ACO architecture with Object (O) Layer, Object Abstraction (OA) Layer, Virtual Object (VO) Layer, Cloud Services (CS) Layer, and Applications Layer. We used these layers and the communication among different entities to propose access control framework. In the enhanced ACO architecture, we identified two modes of interaction between any two layers: direct interaction (DI) and indirect interaction (IdI). DI represents interaction within this layer and between its immediate adjacent layers; and IdI represents interaction with second level of adjacent layers above and below that layer. We have defined the authorization framework categories and ITS communication scenarios. Real-world use cases with single and multi-cloud scenarios and access control requirements reflect the need and use of authorization framework for ITS and connected cars ecosystem. This enabled us to discuss some access control models and authorization approaches relevant for wearable and vehicular IoT ecosystem which have different needs and access control requirements.

8.1.3   Access Control Formal Models

Based on the access control architectures and framework defined in the earlier chapters, we developed different formal models for the well established and widely used cloud IoT platforms, followed by secure communication for virtual objects and discussed attribute based access and communication control models. We studied two widely used and successful cloud IoT platforms, Amazon Web Services (AWS) IoT and Google Cloud Platform (GCP). To bridge the gap between academic research and industry IoT deployments, we developed formal access control models for AWS, referred to as AWSAC, and incorporated IoT specific components to define AWS-IoTAC access control model. Similarly, for GCP, we studied the formal access control model (GCPAC), and then extend this model to a formal Google Cloud Platform IoT Access Control (GCP-IoTAC) model with IoT abstractions. We also illustrated some IoT and CPS use case scenarios to define policies supported in these IoT platforms. Next we discussed access control models for Virtual Objects (VO) communication, and the administrative access control for VO communication. We showed the general access control model for the Amazon Web Services for IoT called AWS-IoTAC. We presented the AWS-IoT access control model for virtual objects called AWS-IoT-ACMVO which is an extension of AWS-IoTAC. Two use cases were discussed to illustrate the communication among virtual objects in the AWS and how this communication can be controlled. We presented first scenario as a simple use case of sensing the speed of single car and the second scenario as a use case of sensing the speed of multiple cars. In the final chapter, we proposed a formal dynamic groups and attribute-based access control (ABAC) model, called CV-ABAC$_G$, for cloud assisted connected cars and intelligent transportation that captures the attribute based fine grained security policies together with user privacy preferences to decide on which notifications or alerts to receive and how to act upon them. In addition, we also proposed a novel perspective with Attribute-Based Communication Control (ABCC) model to secure communications and data flow in IoT and CPS ecosystem. We further analyzed the applicability of ABAC-CC in different IoT domains.

8.2 Future Research Directions

In this book, we primarily focused on proposing access control models and different architectures which can be applied and extended to satisfy the needs of distributed and dynamic IoT and CPS ecosystem. We provide different perspective to understand the access control needs in these systems which is different and unique then the conventional enterprise like single administrator systems with relatively static environment. We consider this brief as a guidebook to foster new research ideas and implementation approaches for the growing IoT dependent and connected world.

In this brief, we proposed different *operational* access control models in IoT and CPS, however, an open question still remains for the understanding and development of *administrative* models which specify facilities to create entities and abstractions, define conditions, contextual constraints and other factors which help in operational model. It is a future research goal to have a consensus in defining an administrative model which can fit into different CPS domains and can be administered by various entities in the distributed systems. In addition, feasible enforcement architectures to identify cloud, edge or hybrid architectures must be supported for the deployment of the operational and administrative models. Federated systems are also needed in CPS domains which involve collaboration and integration of multiple entities, and may need to control and read data from sensors and smart devices in outside domains. This needs the development of cross domain interactions. Further, policy languages must be developed to express the policies which are flexible and extensible to different CPS and IoT domains. The policy language must be able to specify different abstractions and capture different types of constraints which are defined in the access control model.

Convergence of access control models is also needed to develop *hybrid* and *crossbreed* which will enable different access control models to synergistically converge at both policy and enforcement layers. New access control principles are needed to be defined to support expressive policies for CPS domains. Automated security mechanisms are needed in the future AI driven world, and it is important to develop access control solutions which can automatically define security policies based on the event logs and other system audit files.

As the world moves and envision novel IoT and CPS applications, new access control requirements will be needed. This will further answer the question if to adapt existing and established access control models, or create next generation models for the future connected systems.

8.2.1 Communication Control in IoT and CPS

With the proposed ABAC-CC framework in Chap. 7, the goal is to develop secure and user-privacy enhanced access and communication control for CE-IoT architectures. In order to achieve this goal of security and privacy enhanced CE-IoT and enabling secure future smart communities, here we discuss some potential future research directions for securing communication and data flow in IoT and CPS.

- **Artificial Intelligence and Machine Learning:** With a vision of future smart communities enabled by intelligent and autonomous systems, AI and ML are key technologies which require significant research to enable secure communication and data flow between autonomous entities and devices. Thus, further research on AI and ML techniques for enhancing security and privacy in IoT and CPS is essential.
- **Distributed Computing**: There is a need for detailed research on areas of distributed computing infrastructure and technologies, such as Blockchain-based trust frameworks, and distributed and dynamic access control and communication control models, and robust communication and networking protocols along with various cloud computing services.
- **Collaborative IoT Models:** There are various IoT application domains and still new domains are emerging constantly based on IoT devices and data-driven applications. For a sustainable growth of IoT, collaboration among various components, cloud computing platforms, and edge cloudlets is inevitable. It is necessary to establish secure and trustworthy collaboration models based on attributes of entities and relationship between these entities in the IoT space.
- **Insider Threats and Rogue Devices:** ABCC model enables users to define privacy preserving communication control policies. However, even if there are user-centric communication control policies defined for ensuring user data privacy, what if the user data is corrupted or manipulated by an attacker, or if the IoT devices are comprised? Significant research on these aspects is needed to develop appropriate defense mechanisms considering the attackers' actions, such as insider threat or unauthorized physical access gained to the IoT devices. An attacker would turn users' devices into IoT Bots. These scenarios demand more research and enable users to securely access and communicate with their IoT devices and defend as well as prevent against such attacks.

- **Dynamic Edge and Fog Computing:** Edge computing has become a part of the CE-IoT architecture. The smart devices and sensors at edge network and users are continuously moving along carrying highly sensitive data, especially in IoT domains such as Internet of Vehicles (IoV), Wearable IoT, and Internet of Battlefield Things (IoBT). The edge and/or fog computing infrastructure is still evolving and requires further research for supporting various futuristic IoT application domains.

Index

A
Access control, 2, 19, 39, 63, 97, 126, 148, 165
  formal models, 68–71, 132, 166–167
  framework, vi, 8, 19, 34, 36, 39–60, 97,
    122, 148–150, 152, 155, 159–162, 165,
    166, 168
Access control list (ACL), 7, 8, 51, 52, 79,
    97–106, 108, 111
Access control oriented (ACO), 19–36, 39–44,
    46, 48, 58, 64, 71, 72, 97, 98, 111–113,
    122, 150–152, 159, 165, 166
Administrative model, 6, 44, 97–99, 103,
    167
Amazon Web Services (AWS), 8, 23, 43, 63,
    97, 126, 153, 166
Amazon Web Services Access Control
    (AWSAC), 64–68, 71, 94, 166
Amazon Web Services for Internet of Things
    (AWS-IoT), 8, 34, 43, 64, 97, 126, 153,
    166
Application, 1, 19, 39, 67, 97, 125, 148,
    165
Architectures
  access control oriented (ACO), 19–36,
    39–44, 46, 64, 71, 72, 97, 122, 150–152,
    165, 166
  extended access control oriented (E-ACO),
    19, 29–32, 35, 36, 48–50, 165
Artificial intelligence (AI), 1, 10, 23, 147, 148,
    152, 167, 168
Attribute-based access and communication
    control (ABAC-CC), 149, 150, 155,
    159–162, 166–168

Attribute-based access control (ABAC), 7, 8,
    12, 26, 51, 52, 57, 64, 89–94, 97–99,
    101–103, 105–108, 111, 125–144,
    154–156, 158, 159, 161, 162, 167
Attribute-based communication control
    (ABCC), 12, 149, 151, 155–162, 167,
    168
AWS-IoT Access Control (AWS-IoTAC), 52,
    64, 65, 67–76, 89–91, 94, 166

B
Basic safety message (BSM), 30, 35, 39, 46,
    50, 53, 56, 125, 127, 130

C
Capability based access control (CapBAC), 8,
    12, 51, 52
Challenges, 2, 3, 12–14, 39, 91, 94, 147, 148,
    162
Cloud (CL), 1, 19, 39, 63, 97, 125, 147, 165
Cloud computing, 9–11, 24, 28, 91, 148, 159,
    168
Cloud-Enabled Internet of Things (CE-IoT),
    10, 19, 34, 40, 44, 67, 147–154,
    158–162, 166, 168, 169
Cloud-Enabled Wearable Internet of Things
    (CE-WIoT), 40–44, 166
Cloud service (CSCSR), 1, 11, 12, 24, 26,
    27, 29, 30, 32–34, 36, 39–41, 43, 46,
    49–51, 56, 57, 63–67, 71–73, 76–80,
    84, 85, 94, 97, 122, 126, 148–150, 158,
    159, 166

© The Author(s), under exclusive license to Springer Nature Switzerland AG 2022   171
M. Gupta et al., *Access Control Models and Architectures For IoT and Cyber
Physical Systems*, https://doi.org/10.1007/978-3-030-81089-4

Clustered object (CO), 19, 20, 22, 29–32, 35, 36, 39, 40, 47, 49–51, 53, 56, 70, 129–135, 165
Communication control, 12, 32, 41–43, 67, 98, 122, 147–162, 166, 168–169
Connected vehicles (CV), 22, 34, 35, 39, 51, 54, 56, 125–128, 130
Constrained application protocol (CoAP), 32, 51, 148
Controller area network (CAN), 35, 39, 50–52, 59, 127
Cyber physical systems (CPS), v, vi, 1–14, 19–36, 39, 41, 46, 51, 52, 54–60, 63, 122, 125, 126, 147–162, 165–169

D
Data distribution service (DDS), 51
Dedicated short range communication (DSRC), 30, 35, 50, 56, 125, 127
Deep learning (DL), (Found only in Acronyms)
Department of Motor Vehicles (DMV), 57, 58
Direct interaction (DI), 42, 43, 166
Discretionary access control (DAC), v, 7
Dynamic groups, 52, 126–137, 144, 167

E
Edge computing, 1, 9–10, 12, 28, 29, 32, 41, 142, 147, 150–153, 160, 162, 169
Electronic control unit (ECU), 22, 29, 35, 36, 50, 56, 59, 125, 127, 131
Extended access control oriented (E-ACO), 19, 29, 30, 32, 35, 36, 49, 50, 56, 60, 165

F
Fog (FG), 10, 13, 23, 36, 51, 53, 54, 56, 57, 169

G
Gateway, 3, 10, 11, 19, 20, 28–29, 32, 34–36, 39, 41, 43, 44, 46, 51, 56, 57, 59, 149, 150, 152–154, 157–161, 165
Google Cloud Platform (GCP), vi, 63, 64, 67, 76–89, 91–94, 166
Google Cloud Platform Access Control (GCPAC), 64, 76–89, 92–94, 166
Google Cloud Platform IoT Access Control (GCP-IoTAC), 64, 80, 82, 85, 89, 91–94, 166

H
High Impact and Low Frequency (HILF), 48
Hypertext Transfer Protocol (HTTP), 32, 70, 80, 83

I
Indirect interaction (IdI), 42, 43, 51, 166
Intelligent transportation, 20, 22, 34, 39, 46, 47, 125–144, 165, 167
Intelligent transportation system (ITS), 34–36, 48, 50–60, 127, 128, 130, 132, 143, 165, 166
Internet of Things (IoT), v, vi, 1–14, 19–36, 39–46, 52, 58–60, 63–94, 97, 98, 103, 108–111, 116, 122, 125, 126, 147–162, 165–169
Internet of Vehicles (IoV), 11, 19, 22, 34, 169

J
JavaScript Object Notation (JSON), 9, 66, 71, 75, 83, 89, 109, 110, 157

M
Machine Learning (ML), 1, 10, 23, 64, 79, 147, 148, 153, 168
Machine-to-Machine (M2M), 10, 68
Mandatory Access Control (MAC), v, 7
Medical IoT (MIoT), 21
Message Queuing Telemetry Transport (MQTT), 9, 32, 34, 51, 52, 56, 63, 68, 70, 73, 80, 83–85, 89, 90, 97, 99, 108–110, 114–117, 119, 138, 141, 148

O
Object (OB), 1, 19, 39, 64, 97, 126, 150, 165
Object abstraction (OA), 19, 28, 32, 33, 40–44, 46, 92, 93, 151, 155, 165, 166
Object Layer Application (OAP), 50
On Board Diagnostic (OBD), 52, 59
Operational model, 44, 53, 97–105, 107, 111, 167
Over the Air (OTA), 36, 48, 56, 57, 59, 92, 93, 127

P

Protocol, v, 1, 6, 9, 10, 12, 28, 32, 34, 35,
    50–52, 56, 63, 68, 70, 73, 80, 83–85,
    90, 125, 127, 147, 148, 168
Public key infrastructure (PKI), 26, 53

R

Radio frequency identification (RFID), 3, 115,
    116, 120, 121
Relationship-based access control (ReBAC), 8,
    26
Remote patient monitoring (RPM), 32–34,
    44–46, 48
Role-based access control (RBAC), v, 7, 8, 12,
    26, 63–67, 78, 79, 91–93, 97–99, 103,
    105–108, 148, 153, 161

S

Security Credential and Management System
    (SCMS), 53
Smart cars, 13, 19–23, 30, 35, 46, 47, 49–52,
    54, 55, 57, 58, 125–128, 130, 165

T

Taxonomy, 20–23, 36
Telematic control unit (TCU), 35, 51

Tire pressure monitoring (TPS), 22, 30, 35, 56
Topic-based, 2, 9, 14, 51, 98, 112

U

United States Department of Transportation
    (USDOT), 53
User (U), 2, 20, 39, 65, 99, 125, 147, 167

V

Vehicle-to-everything (V2X), 48, 53, 56
Vehicle-to-infrastructure (V2I), 47, 48, 53, 56,
    125, 127, 130
Vehicle-to-pedestrian (V2P), 48, 56
Vehicle-to-vehicle (V2V), 35, 47, 52, 53, 56,
    125, 127, 130
Vehicular cloud (VC), 24, 36, 54, 57
Vehicular IoT (VIoT), 13, 19, 21, 30, 34, 39,
    40, 151, 160, 166
Virtual objects (VOVOB), 5, 19, 39, 68, 97,
    138, 149, 165

W

Wearable IoT (WIoT), 13, 14, 19–21, 28, 29,
    36, 39–46, 59, 150, 153, 157, 158, 165,
    166, 169
Wireless sensor network (WSN), 21, 46

Printed in the United States
by Baker & Taylor Publisher Services